Acknowledgements

Show What You Know® Publishing acknowledges the following for their efforts in making this assessment material available for Florida students, parents, and teachers.

Cindi Englefield, President/Publisher
Eloise Boehm-Sasala, Vice President/Managing Editor
Jill Borish, Production Editor
Christine Filippetti, Production Editor
Charles V. Jackson, Mathematics Editor
Jennifer Harney, Editor/Illustrator
Lainie Burke Rosenthal, Editor/Graphic Designer
Melissa Blevins, Assistant Editor
Angela Gorter, Assistant Editor

About the Contributors

The content of this book was written BY teachers FOR teachers and students and was designed specifically for the Florida Comprehensive Assessment Test (FCAT) for Grade 7 Mathematics. Contributions to the Mathematics section of this book were also made by the educational publishing staff at Show What You Know® Publishing. Dr. Jolie S. Brams, a clinical child and family psychologist, is the contributing author of the Test Anxiety and Test-Taking Strategies chapters of this book. Without the contributions of these people, this book would not be possible.

Table of Contents

Introduction

About the Show What You Know® Program

Show What You Know® Publishing has been developing test-preparation products since 1993. These products teach test-taking skills that are specific to the Florida Comprehensive Achievement Tests and provide students with practice on full-length tests that simulate the format of the FCAT. We understand that many students are not good test takers because they get nervous and may experience test anxiety. To help students with this issue, we provide a chapter, written by a child psychologist, that explains what test anxiety is, what it feels like, and ways to reduce the anxiety before, during, and after the testing situation has occurred. Research has proven that three elements must exist for test success: knowledge, test-taking skills, and confidence. The Show What You Know® test-preparation program helps to ensure test success.

How to Use This Program

There are numerous ways to use the Student Workbook and Parent/Teacher Supplement to help prepare your students for the Mathematics FCAT. But before your students begin the Mathematics Tutorial and the Practice Tests, take time to walk them through the Test Anxiety and Test-Taking Strategies chapters to help them learn how to be a better test taker. These important concepts can really improve students' test scores—it works!

Tips to Reduce Test Anxiety

You can read the entire chapter on test anxiety, or you can break it up into different sections, allowing yourself plenty of time to discuss and practice the methods suggested to reduce test stress. Identify what test anxiety is, and help students learn the symptoms (see Student Workbook, page 2). Explain to students that test anxiety is normal and that there are ways it can be overcome. Tell them how being a little nervous will motivate them to do their best, but being very nervous could make them forget information they need to know for the test.

Pages 4 and 5 of the Student Workbook identify different ways that test anxiety can manifest itself in students. Have your students read about these characteristics and see if they can identify with any of them. Again, explain to them that it's okay to feel anxious about tests, and reassure them that many students feel test stress.

The rest of the Test Anxiety chapter offers activities to show ways to overcome test stress, such as thinking positively instead of negatively, emphasizing the importance of good physical health, and studying and practicing for the test. These are skills that will help your students succeed on the Mathematics FCAT, as well as other tests they will face throughout their lives.

Test-Taking Strategies That Work!

The Test-Taking Strategies chapter can be used in a class discussion. Ask for student volunteers to read a strategy out loud, then discuss as a class what the strategy is teaching. The strategies listed in the chapter include:

- Be an active learner
- Don't depend on luck
- Do your best everyday
- Get to know the FCAT
- Read directions and questions carefully
- Know how to fill in the answer bubbles
- Don't speed through the test
- Answer every question
- If you don't know the answer, guess
- Don't get stuck on one question
- Always recheck your work
- Use the FCAT Mathematics Reference Sheet
- Pay attention to yourself and not others

After reviewing each strategy, ask students if they have found ways to help them relax and prepare for tests. Maybe they like to read a book, listen to music, or they get extra sleep the week of a test. All of these strategies can be used to help them do well on the Mathematics FCAT and other tests.

The Purpose of the Tutorial

After you have walked your students through the Test Anxiety and Test-Taking Strategies chapters, begin with the Mathematics Tutorial. The Tutorial is designed to walk students through each Standard, Benchmark, and Grade-Level Expectation that may be assessed by the Mathematics FCAT for Grade 7. A sample question is provided for each Standard, Benchmark, and Grade-Level Expectation. An Answer Sheet is located at the end of the Tutorial to help students become more familiar with answering questions in the space provided. A complete analysis is given in the Supplement to explain to the student why the correct answer is correct, and why the incorrect answers are incorrect. The purpose of the Tutorial is to help students become more familiar with the standards they may be tested on and to become familiar with the types of questions that they could see on the actual test for each standard.

The Practice Tests

Once students have completed the Tutorial, they can take the Assessment. Each Assessment has 50 questions, and the Item Distribution matches what they will see on the actual Mathematics FCAT for Grade 7. These tests were designed to simulate the Mathematics FCAT so that students can become familiar with the actual look of the test. The more familiar they are with the look of the test, the more confidence they will have when they take the actual Mathematics FCAT for Grade 7. An Answer Document for each Assessment is provided in the Student Workbook. Assessment One can be used as a pre-test to find your students' strengths and weaknesses, and which standards need additional review. A Correlation Chart is provided to assist you with this. Assessment Two can be given as a post-test after reviewing the standards identified from Assessment One that needed more attention.

Correlation Charts Track Students' Strengths and Weaknesses

In this Parent/Teacher Supplement, there are Correlation Charts for each Assessment. The Standard, Benchmark, Grade-Level Expectation, and answers are listed for each question. To use the chart, write the students' names in the left-hand column. When students miss a question, place an "X" in the corresponding box. A column with a large number of "Xs" shows that your class needs more practice with that particular standard. You can quickly identify the needs of individual students.

Answer Key with Analysis and Feedback

The question analyses with answers are provided in the Parent/Teacher Supplement. The question from the assessment is repeated, the correct answer is given, and the analyses explain in detail why each question is correct.

Additional Teaching Tools

A Glossary of Mathematics Terms is provided to help your students understand terms that they should be familiar with before taking the Mathematics FCAT for Grade 7. General teaching tips are also provided to give you suggestions on how to incorporate the Sunshine State Standards into your curriculum.

More Test-Preparation Products

Additional test preparation products are available for your use. The Show What You Know® on the FCAT for Grades 6–8 Mathematics Flash Cards provide 96 additional mathematics questions. Each question card has the question on the front and the answer with an analysis on the back. The analysis explains why correct answers are correct and why incorrect answers are incorrect. These flash cards can be used as Questions of the Day, for in-class quizzes, for games, in group projects, and as homework.

Suggested Timeline for Program Use

Now that you have a better understanding of how to use this Show What You Know® Test-Preparation Program, here is a suggested timeline for you to use to incorporate the products into your teaching schedule.

Suggested Timeline for Using the Show What You Know® on the FCAT for Grade 7 Mathematics, Student Workbook	
Week 1	Test Anxiety chapter
Week 2	Test-Taking Strategies chapter
Week 3	Mathematics Practice Tutorial
Week 4	Mathematics Assessment One
Week 5	Additional Review after results of Assessment One
Week 6	Additional Review after results of Assessment One
Week 7	Mathematics Assessment Two
Week 8	Additonal Review after results of Assessment Two

Thank you for implementing the Show What You Know® Test-Preparation Program in your classroom. Good luck to you and all of your students as they prepare for the FCAT!

Understanding Grade Level Expectations

Subject Area: MA: Mathematics

A **strand** is a category of knowledge. The five strands assessed on the FCAT Mathematics test are (A) Number Sense, Concepts, and Operations, (B) Measurement, (C) Geometry and Spatial Sense, (D) Algebraic Thinking, and (E) Data Analysis and Probability.

Each Mathematics **standard** is a general statement of expected student achievement within a strand. The standards are the same for all grade levels.

Benchmarks are specific statements of expected student achievement under each Mathematics standard. Test items are written to assess the benchmarks. In some cases, two or more related benchmarks are grouped together because the assessment of one benchmark necessarily addresses another benchmark.

> **MA: Mathematics**
>
> **Strand A: Number Sense, Concepts, and Operations**
>
> **Standard 1:**
> **The student understands the different ways numbers are represented and used in the real world.**
>
> 1. Associates verbal names, written word names, and standard numerals with integers, fractions, decimals; numbers expressed as percents; numbers with exponents; numbers in scientific notation; radicals; absolute value; and ratios.

The Grade Level Expectation's **Numbering System** identifies the Subject Area, the Strand, the Level, and the Benchmark. For example, in the Grade Level Expectation (MA.A.1.3.1), the first letters (MA) stand for the Subject Area, Mathematics, and the second letter (A.) for the Strand, (Number Sense). The first number (1.) stands for the Standard, the second number (3.) for the Level band (Grades 6–8), the third number (1.) for the Benchmark. **Note:** The Grade Level Expectations are not intended to take the place of a curriculum guide, but rather to serve as the basis for curriculum development to ensure that the curriculum is rich in content and is delivered through effective instructional activities. The Grade Level Expectations are in no way intended to limit learning, but rather to ensure that all students across the state receive a good educational foundation that will prepare them for a productive life.

Sunshine State Standards

Mathematics

Strand A: Number Sense, Concepts, and Operations

Standard 1: The student understands the different ways numbers are represented and used in the real world.

MA.A.1.3.1 Associates verbal names, written word names, and standard numerals with integers, fractions, decimals; numbers expressed as percents; numbers with exponents; numbers in scientific notation; radicals; absolute value; and ratios. **(MC, GR)**

MA.A.1.3.2 Understands the relative size of integers, fractions, and decimals; numbers expressed as percents; numbers with exponents; numbers in scientific notation; radicals; absolute value; and ratios. **(MC)**

MA.A.1.3.3 Understands concrete and symbolic representations of rational numbers and irrational numbers in real-world situations. **(MC, GR)**

MA.A.1.3.4 Understands that numbers can be represented in a variety of equivalent forms, including integers, fractions, decimals, percents, scientific notation, exponents, radicals, and absolute value. (Also assesses A.1.3.1 and A.1.3.3) **(MC, GR)**

Standard 2: The student understands number systems.

MA.A.2.3.1 Understands and uses exponential and scientific notation. **(MC, GR)**

Standard 3: The student understands the effects of operations on numbers and the relationships among these operations, selects appropriate operations, and computes for problem solving.

MA.A.3.3.1 Understands and explains the effects of addition, subtraction, multiplication, and division on whole numbers and fractions, including mixed numbers and decimals, including the inverse relationships of positive and negative numbers. **(MC)**

MA.A.3.3.2 Selects the appropriate operation to solve problems involving addition, subtraction, multiplication, and division of rational numbers, ratios, proportions, and percents, including the appropriate application of the algebraic order of operations. **(MC, GR)**

MA.A.3.3.3 Adds, subtracts, multiplies, and divides whole numbers, decimals, and fractions, including mixed numbers, to solve real-world problems, using appropriate methods of computing, such as mental mathematics, paper and pencil, and calculator. **(MC, GR)**

(MC=Multiple Choice; GR=Gridded Response)

Standard 4: The student uses estimation in problem solving and computation.

MA.A.4.3.1 Uses estimation strategies to predict results and to check the reasonableness of results. (Also assesses A.4.2.1, B.2.3.1, and B.3.3.1) **(MC)**

Standard 5: The student understands and applies theories related to numbers.

MA.A.5.3.1 Uses concepts about numbers, including primes, factors, and multiples, to build number sequences. **(MC, GR)**

Strand B: Measurement
Standard 1: The student measures quantities in the real world and uses the measures to solve problems.

MA.B.1.3.1 Uses concrete and graphic models to derive formulas for finding perimeter, area, surface area, circumference, and volume of two- and three-dimensional shapes, including rectangular solids and cylinders. (Also assesses B.1.2.2 and B.2.3.1) **(MC, GR)**

MA.B.1.3.2 Uses concrete and graphic models to derive formulas for finding rates, distance, time, and angle measures. (Also assesses B.1.2.2 and B.2.3.1) **(MC)**

MA.B.1.3.3 Understands and describes how the change of a figure in such dimensions as length, width, height, or radius affects its other measurements such as perimeter, area, surface area, and volume. (Also assesses C.2.3.1) **(MC, GR)**

MA.B.1.3.4 Constructs, interprets, and uses scale drawings such as those based on number lines and maps to solve real-world problems. (Also assesses B.2.3.1) **(MC, GR)**

Standard 2: The student compares, contrasts, and converts within systems of measurement (both standard/nonstandard and metric/customary).

MA.B.2.3.1 Uses direct (measured) and indirect (not measured) measures to compare a given characteristic in either metric or customary units. **(MC, GR)**

MA.B.2.3.2 Solves problems involving units of measure and converts answers to a larger or smaller unit within either the metric or customary system. **(MC, GR)**

Standard 3: The student estimates measurements in real-world problem situations.

MA.B.3.3.1 Solves real-world and mathematical problems involving estimates of measurements including length, time, weight/mass, temperature, money, perimeter, area, and volume, in either customary or metric units. **(MC)**

Strand C: Geometry and Spatial Sense

Standard 1: The student describes, draws, identifies, and analyzes two- and three-dimensional shapes.

MA.C.1.3.1 Understands the basic properties of, and relationships pertaining to, regular and irregular geometric shapes in two and three dimensions. (Also assesses C.1.2.1) **(MC)**

Standard 2: The student visualizes and illustrates ways in which shapes can be combined, subdivided, and changed.

MA.C.2.3.1 Understands the geometric concepts of symmetry, reflections, congruency, similarity, perpendicularity, parallelism, and transformations, including flips (reflections), slides (translations), turns (rotations), and enlargements (dilations). (Also assesses B.1.3.3, C.1.2.1, C1.3.1, and C.3.3.1) **(MC)**

MA.C.2.3.2 Predicts and verifies patterns involving tessellations (a covering of a plane with congruent copies of the same pattern with no holes and no overlaps, like floor tiles). **(MC)**

Standard 3: The student uses coordinate geometry to locate objects in both two and three dimensions and to describe objects algebraically.

MA.C.3.3.1 Represents and applies geometric properties and relationships to solve real-world and mathematical problems. (Also assesses C.2.3.1, C.2.3.2, and C.3.2.2) **(MC)**

MA.C.3.3.2 Identifies and plots ordered pairs in all four quadrants of a rectangular coordinate system (graph) and applies simple properties of lines. **(MC)**

Strand D: Algebraic Thinking

Standard 1: The student describes, analyzes, and generalizes a wide variety of patterns, relations, and functions.

MA.D.1.3.1 Describes a wide variety of patterns, relationships, and functions through models, such as manipulatives, tables, graphs, expressions, equations, and inequalities. (Also assesses A.5.3.1) **(MC, GR)**

MA.D.1.3.2 Creates and interprets tables, graphs, equations, and verbal descriptions to explain cause-and-effect relationships.(Also assesses A.5.3.1) **(MC, GR)**

Standard 2: The student uses expressions, equations, inequalities, graphs, and formulas to represent and interpret situations.

MA.D.2.3.1 Represents and solves real-world problems graphically, with algebraic expressions, equations, and inequalities. (Also assesses A.1.3.3) **(MC)**

MA.D.2.3.2 Uses algebraic problem-solving strategies to solve real-world problems involving linear equations and inequalities. **(MC, GR)**

Strand E: Data Analysis and Probability

Standard 1: The student understands and uses the tools of data analysis for managing information.

MA.E.1.3.1 Collects, organizes, and displays data in a variety of forms, including tables, line graphs, charts, and bar graphs, to determine how different ways of presenting data can lead to different interpretations. (Also assesses E.1.3.3) **(MC, GR)**

MA.E.1.3.2 Understands and applies the concepts of range and central tendency (mean, median, and mode). (Also assesses E.1.3.3) **(MC, GR)**

MA.E.1.3.3 Analyzes real-world data by applying appropriate formulas for measures of central tendency and organizing data in a quality display, using appropriate technology, including calculators and computers. **(MC, GR)**

Standard 2: The student identifies patterns and makes predictions from an orderly display of data using concepts of probability and statistics.

MA.E.2.3.1 Compares experimental results with mathematical expectations of probabilities. **(MC)**

MA.E.2.3.2 Determines odds for and odds against a given situation. (Also assesses E.2.2.2) **(MC)**

Standard 3: The student uses statistical methods to make inferences and valid arguments about real-world situations.

MA.E.3.3.1 Formulates hypotheses, designs experiments, collects and interprets data, and evaluates hypotheses by making inferences and drawing conclusions based on statistics (range, mean, median, and mode) and tables, graphs, and charts. (Also assesses E.3.3.2) **(MC)**

MA.E.3.3.2 Identifies the common uses and misuses of probability and statistical analysis in the everyday world. **(MC)**

Mathematic Terms (complete Glossary of Mathematic Terms can be found starting on page 39 of the Show What You Know® on the 7th Grade FCAT, Mathematics Student Workbook)

absolute value
acute angle
acute triangle
addend
addition
additive identity
additive inverse property
algebraic equation (or inequality)
algebraic expression
algebraic rule
altitude
analyze
angle
approximate
approximation
area
argument
associative property
attribute
average
axes
bar graph
base
box-and-whisker plot
break
capacity
central angle
chart
circle
circle graph
circumference
closed figure
cluster
combination
common denominator
common multiple
commutative property
compare
complementary angles
complementary events
composite number
compound probability
conclude
conclusion
cone
congruent figures
conjecture
constant
contraction
contrast
coordinates
counting principle

cross-multiply
cube
customary system
cylinder
data
data displays/graphs
decimal number
denominator
dependent variable
diagonal
diagram
diameter
difference
dilation
dimensions
direct measure
direct proportion
distributive property
dividend
divisible
division
divisor
edge
empirical probability
enlargement
equality
equally likely
equation
equiangular
equilateral
equilateral triangle
equivalent expressions
equivalent forms of a number
estimate
estimation
evaluate an algebraic expression
even number
event
expanded form
experimental probability
exponent
expression
extraneous information
extrapolate
face
factor
figure
flip
fraction
function
function machine
function table

graph
greatest common factor (divisor)
grid
height
hexagon
histogram
hypotenuse
hypothesis
identity property
independent variable
indirect measure
improper fraction
independent events
inequality
integer
intercept
interpret
intersecting lines
interval
inverse property
inverse operation
irrational numbers
isosceles triangle
justify
labels (for a graph)
least common denominator
least common multiple (LCM)
length
likelihood
line
line graph
line segment
line of symmetry
line plot
linear equation
linear inequality
mass
mean
median
method
metric system
midpoint
mixed number
mode
multiple
multiplication
multiplicative identity
multiplicative inverse (reciprocal)
mutually exclusive
natural numbers
negative exponent
net (of a three-dimensional shape)

nonstandard units of measure
number line
number sentence
numerator
obtuse angle
obtuse triangle
octagon
odd number
odds (for and against)
open figure
operation
operation symbol
operational shortcut
order of operations
ordered pairs
organized data
origin
outcome
outlier
parallel
parallelogram
pattern
pentagon
percent
perimeter
permutation
perpendicular line
pi
pictograph
pie chart
place value
plane
plane figure
plot
point
polygon
polyhedron
population
power
precision
predict
prediction
prime numbers
prism
probability
product
proper fraction
properties
proportion
proportional
pyramid
Pythagorean theorem
quadrant

quadrilateral
questionnaire
quotient
radical
radicand
radius
randomly (chosen)
range
rate
ratio
rational number
ray
real number
reasonable
reciprocal
rectangle
reduce
reduction
reflection
regular polygon
relation
relative size
represent
rhombus
right angle
right circular cylinder
right prism or rectangular solid
right triangle
rise
rotation
rounding
rule
ruler
run
sample
sample space
scale
scale factor
scale model
scalene triangle
scatterplot
scientific notation
segment
semicircle
sequence
set
side
similar
simplify
slide
slope
solid figures
solve

sphere
square
square number
square root
squiggle
standard units of measure
stem-and-leaf plot
straight angle
strategy
subtraction
successive events
sum
summary
supplementary angle
surface area
survey
symbol
symbolic representations
　　of numbers
symmetrical
table
tessellation
theoretical/expected probability
three-dimensional figure
transformation
translation
transversal
trapezoid
tree diagram
trend
triangle
turn
two-dimensional figure
U.S. system of measurement
undefined terms
unknown
unorganized data
validate
variable
verify
vertex
vertical angles
vertices
volume
whole number
weight
word forms
x-axis
x-intercept
y-axis
y-intercept
zero property

Glossary of Illustrations (Full sized Glossary of Illustrations can be found starting on page 33 of the Show What You Know® on the 7th Grade FCAT, Mathematics Student Workbook)

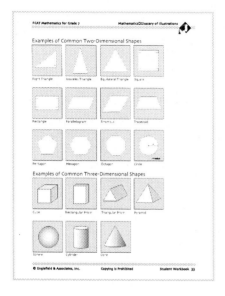

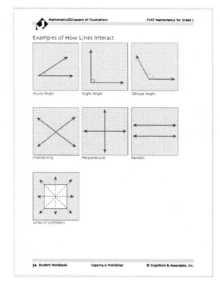

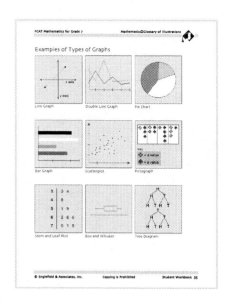

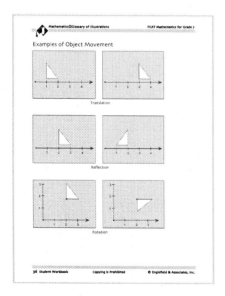

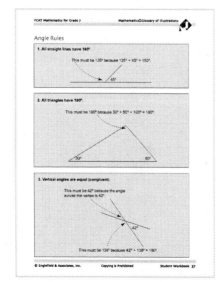

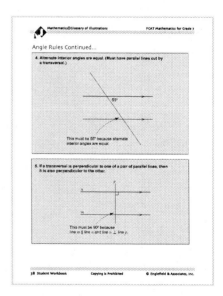

Mathematics Practice Tutorial Introduction Page as it appears on page 65 in the
Show What You Know® on the 7th Grade FCAT, Mathematics Student Workbook.

Mathematics Practice Tutorial

Directions for the Mathematics Practice Tutorial

The Mathematics Practice Tutorial contains 33 practice questions, a Mathematics Reference Sheet (from page 31), and an Answer Sheet. You will mark your answers on the Answer Sheet on pages 105–108 of this workbook. If you don't understand a question, just ask your teacher to explain it to you.

This section will review the Strands, Standards, and Benchmarks used to assess student achievement in the state of Florida. Following the description of each Benchmark, a sample mathematics practice item is given. Each item gives you an idea of how the Benchmark may be assessed. Review these items to increase your familiarity with FCAT-style multiple-choice and gridded-response items. Once you have read through this tutorial section, you will be ready to complete the Mathematics Assessment tests.

About the FCAT Mathematics for Grade 7

Items in this section of the FCAT will test students' ability to perform mathematical tasks in real-world and mathematical situations and will neither require students to define mathematical terminology nor to memorize specific facts. Seventh-grade students are also allowed to use the Mathematics Reference Sheet during testing. The FCAT is meant to gauge a student's ability to apply mathematical concepts to a given situation.

The FCAT Mathematics will ask students multiple-choice items and gridded-response items. For more in-depth information on how to answer these different types of questions please refer to the Test-Taking Strategies chapter on pages 11–22 of the *Show What You Know® on the 7th Grade FCAT, Mathematics Student Workbook*.

For multiple-choice items, each question has only one correct answer; the other three choices are distractors representing incorrect answers that students commonly obtain for the question. Multiple-choice items are worth one point each. Students should spend no more than one minute answering each individual question, but they should be sure to allow themselves time to scrutinize each possible choice.

Gridded-response items also have only one correct answer, but in certain circumstances, the answer may be represented in different formats. For example, if a question asks what fraction is equal to 50%, students may respond with 1/2, 4/8, 50/100, or any other fraction equaling 50%. Gridded-response questions are worth one point each.

Additional Hints to Remember for Taking the FCAT Mathematics Test

Here are some hints to help students do their best when they take the FCAT Mathematics test. Students should keep these hints in mind when they answer the sample questions.

- Learn how to answer each kind of question. The FCAT Mathematics test for Grade 7 has two types of questions: multiple-choice and gridded-response.

- Read each question carefully and think about ways to solve the problem before you try to answer the question.

- Answer the questions you are sure about first. If a question seems too difficult, skip it and go back to it later.

- Be sure to fill in the answer bubbles correctly. Do not make any stray marks around answer spaces.

- Think positively. Some problems may seem hard to you, but you may be able to figure out what to do if you read each question carefully.

- When you have finished each problem, reread it to make sure your answer is reasonable.

- Relax. Some people get nervous about tests. It's natural. Just do your best.

Sample Multiple-Choice Item

To help students understand how to answer the test questions, look at the sample test question and Answer Sheet below. It is included to show students what a multiple-choice question in the test is like and how to mark their answers on the Answer Sheet.

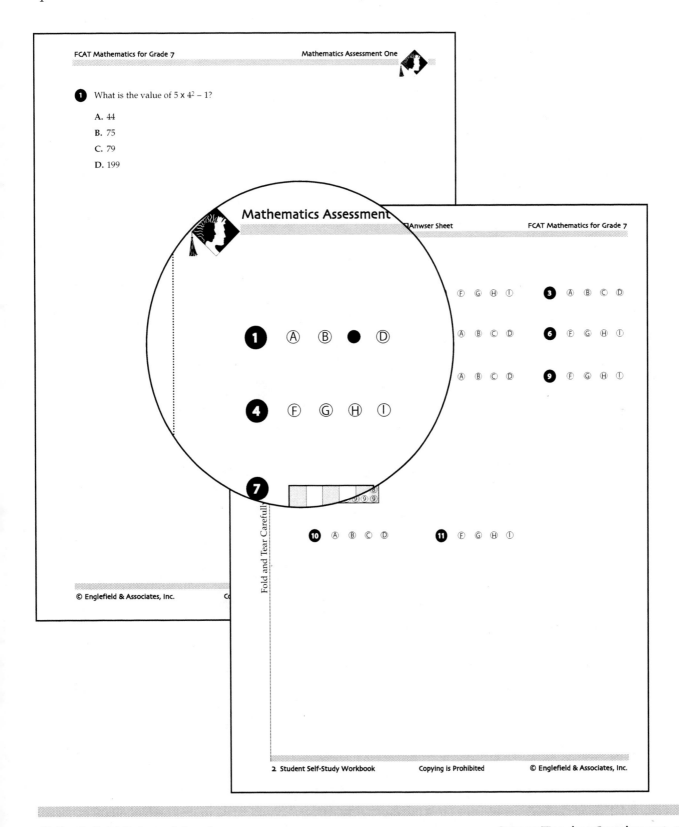

Sample Gridded-Response Item

To help your students understand how to answer the test questions, look at the sample test question and Answer Sheet below. It is included to show students what a gridded-response item in the test is like and how to mark their response on your Answer Sheet.

 Mathematics test questions with this symbol require that students fill in a grid on their answer sheet. There may be more than one correct way to fill in a response grid. The gridded-response section on page 17 will show you the different ways the response grid may be completed.

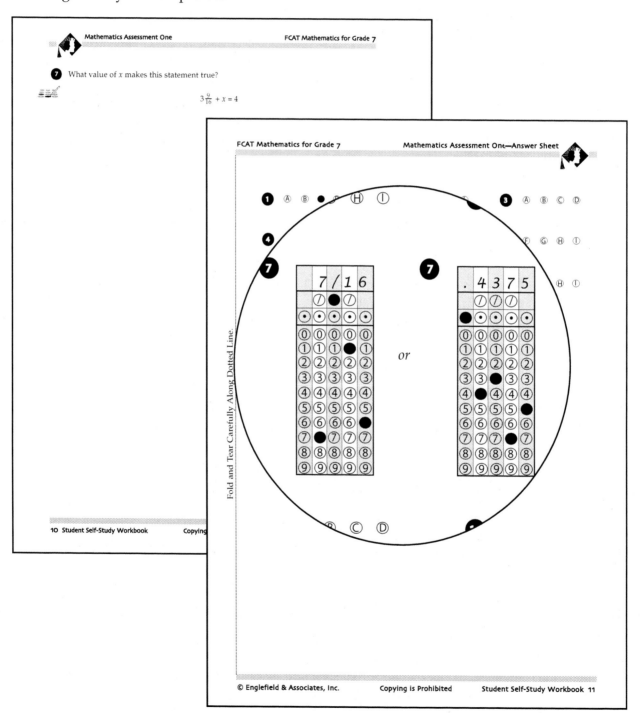

Mathematics Assessment One   FCAT Mathematics for Grade 7

7   What value of x makes this statement true?

$$3\tfrac{9}{16} + x = 4$$

FCAT Mathematics for Grade 7   Mathematics Assessment One—Answer Sheet

or

Fold and Tear Carefully Along Dotted Line

10   Student Self-Study Workbook   Copying

© Englefield & Associates, Inc.   Copying is Prohibited   Student Self-Study Workbook 11

Additional Gridded-Response Item Information

Each gridded-response question requires a numerical answer which should be filled into a bubble grid. The bubble grid consists of 5 columns. Each column contains numbers 0–9 and a decimal point; the middle three columns contain a fraction bar as well. Students do not need to include any commas for numbers greater than 999. When filling in their answer, students should only fill in one bubble per column. All gridded-response questions are constructed so the answer will fit into the grid. Students can print their answer with the first digit in the left answer box, or with the last digit in the right answer box. Print only one digit or symbol in each answer box. Do not leave a blank box in the middle of an answer. Make sure to fill in a bubble under each box in which there is an answer and be sure to write the answer in the grid above the bubbles as well, in case clarification is needed. Answers can be given in whole number, fraction, or decimal form. For questions involving measurements, the unit of measure required for the answer will be provided. When a percent is required to answer a question, do NOT convert the percent to its decimal or fractional equivalent. Grid in the percent value without the % symbol. Students may NOT write a mixed number such as $13\frac{1}{4}$ in the answer grid. If the answer is a mixed number, convert the answer to an improper fraction, such as $\frac{53}{4}$, or to a decimal number, such as 13.25. If students try to fill in $13\frac{1}{4}$, it will be read as $\frac{131}{4}$ and be counted as wrong. Students will also be instructed when to round their answer in a particular way. Some example responses are given below.

Answer: 23,901   Answer: 26.5   Answer: 0.071   Answer: $\frac{3}{8}$

Question **1** *assesses:*

Strand A: Number Sense, Concepts, and Operations

Standard 1: The student understands the different ways numbers are represented and used in the real world.

MA.A.1.3.1 Associates verbal names, written word names, and standard numerals with integers, fractions, decimals; numbers expressed as percents; numbers with exponents; numbers in scientific notation; radicals; absolute value; and ratios. **(MC, GR)**

Teaching Tips

- Make a wall chart similar to the one in the analysis and help students read numbers in sets of three "between the commas" as described in the analysis.

- Develop a transparent place value grid for your students to reuse similar to the sample below. First, using a piece of graph paper, create a place value grid. Using a black marker, create 17 sections: each section should be about 2-4 squares wide and about 20 squares high. Clearly label each section, from hundred millions to hundred millionths. Don't forget to include the decimal point. Copy this grid onto a transparency. As your students are writing out numbers, they can lay the transparency over their papers to understand place value. You may also wish to create a matching answer grid for students to write their numbers in so the overlay fits properly.

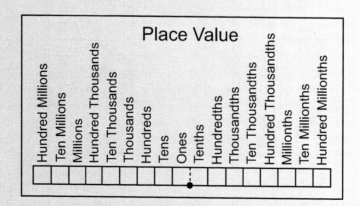

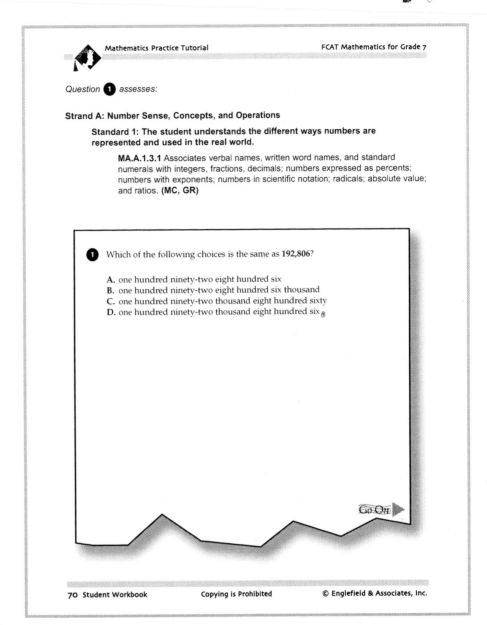

Mathematics Practice Tutorial   FCAT Mathematics for Grade 7

Question **1** *assesses:*

Strand A: Number Sense, Concepts, and Operations

    Standard 1: The student understands the different ways numbers are represented and used in the real world.

        MA.A.1.3.1 Associates verbal names, written word names, and standard numerals with integers, fractions, decimals; numbers expressed as percents; numbers with exponents; numbers in scientific notation; radicals; absolute value; and ratios. **(MC, GR)**

1 Which of the following choices is the same as **192,806**?

    A. one hundred ninety-two eight hundred six
    B. one hundred ninety-two eight hundred six thousand
    C. one hundred ninety-two thousand eight hundred sixty
    D. one hundred ninety-two thousand eight hundred six

Go On ▶

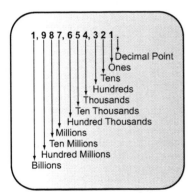

1,9 8 7, 6 5 4, 3 2 1 .
    Decimal Point
    Ones
    Tens
    Hundreds
    Thousands
    Ten Thousands
    Hundred Thousands
    Millions
    Ten Millions
    Hundred Millions
Billions

Analysis:

Choice D is correct. Use the chart above to help read the number in the question. When reading any number always start with the furthest left non-zero digit. It also helps to consider the digits as sets of three "between the commas." For example, in this case, the furthest left digit is a 1 in the hundred thousands place, which is followed by a 9 in the ten thousands place, and a 2 in the thousands place. If you cover up the last three digits, 8, 0, and 6, the remaining number looks like one hundred ninety-two, but since all of the place values are in the thousands, this part of the number is read as one hundred ninety-two thousand. Similarly, if you cover up the first three digits, 1, 9, and 2, the remaining number is read as eight hundred six. Putting both parts together gives us one hundred ninety-two thousand eight hundred six. Choices A and B are incorrect because they make no sense and numbers are not read this way. Choice C is incorrect because one hundred ninety-two thousand eight hundred sixty is written as 192,860, not 192,806.

Question **2** *assesses:*

Strand A: Number Sense, Concepts, and Operations

Standard 1: The student understands the different ways numbers are represented and used in the real world.

MA.A.1.3.2 Understands the relative size of integers, fractions, and decimals; numbers expressed as percents; numbers with exponents; numbers in scientific notation; radicals; absolute value; and ratios. **(MC)**

Teaching Tip

Create a list of 10-20 numbers written on large index cards in a few of these forms: integers, fractions, decimals, percents, numbers with exponents, numbers written in scientific notation, radicals, absolute value, and ratios. Shuffle the deck and ask a student to choose a card and place it on the chalkboard tray in order relative to the other numbers already there. Ask each student to justify his or her placement by changing numbers into the same form and writing the equivalent form on the blackboard above their card. As students become more comfortable with ordering numbers in different forms, increase the list to 20-30 numbers and the variety of forms in the deck.

Question **2** *assesses:*

Strand A: Number Sense, Concepts, and Operations

> **Standard 1: The student understands the different ways numbers are represented and used in the real world.**
>
>> **MA.A.1.3.2** Understands the relative size of integers, fractions, and decimals; numbers expressed as percents; numbers with exponents; numbers in scientific notation; radicals; absolute value; and ratios. **(MC)**

2 Which of the following numbers is the **greatest**?

F. $\frac{190}{10}$

G. 20%

H. 2.2×10^{-1}

I. $\frac{6}{150}$

Go On ▶

Analysis:

Choice F is correct. If you are asked to compare several values in different formats, it's best to convert the numbers to the same format. Many students find it easiest to compare numbers when they are all decimals. Choice F, 190/10, is equivalent to 19. Choice G, 20%, is the same as 0.20. Choice H is the decimal 0.22 in scientific notation and Choice I, 6/150 equals 0.04. Choice F is the only number with a decimal equivalent greater than one, so it is the greatest number. Scientific Notation: A number written in scientific notation has two parts: a decimal fraction between 1 and 10 and some power of 10. To convert a number from scientific to standard notation, you must move the decimal point the same number of spaces as the power of 10. If the power is positive, move the decimal to the right, but if the power is negative, move the decimal to the left. Remember, **a negative exponent does not mean a negative number**, it means a fraction.

Question **3** *assesses:*

Strand A: Number Sense, Concepts, and Operations

Standard 1: The student understands the different ways numbers are represented and used in the real world.

MA.A.1.3.3 Understands concrete and symbolic representations of rational numbers and irrational numbers in real-world situations. **(MC, GR)**

Teaching Tips

Review the following concepts with students:

- Rational numbers can be represented as the ratio of two integers (positive or negative). They can be integers (e.g., 6/3 = 2); fractions (e.g., 5/6); terminating decimals (e.g., 1/4 = 0.25); or repeating decimals (e.g., 13/3 = 4.333).

- Irrational numbers are real numbers that cannot be expressed as the ratio of two integers. Stress to students that these numbers cannot be expressed exactly in our decimal number system. We can get as much precision as we like, but we cannot get an exact decimal answer. A familiar example of an irrational number is pi, π, the ratio of a circle's circumference to its diameter. Pi's value is known to millions of places, but any accepted value is still an approximation. For the purpose of the FCAT, the approximate rational values of 3.14 or 22/7 are accepted. Other examples of irrational numbers include many roots, such as the square root of 2, and e (the base of the natural logarithm). Show students that irrational numbers can be expressed as non-terminating, non-repeating decimals (i.e., the numbers after the decimal point do not repeat in any pattern, but also do not end).

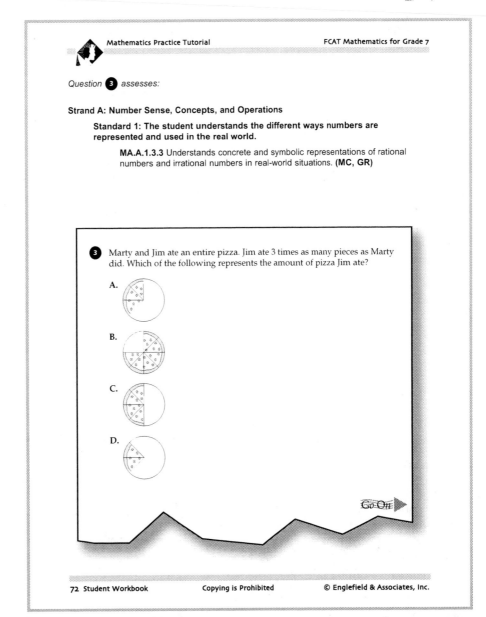

Question **3** assesses:

Strand A: Number Sense, Concepts, and Operations

> **Standard 1: The student understands the different ways numbers are represented and used in the real world.**
>
> > **MA.A.1.3.3** Understands concrete and symbolic representations of rational numbers and irrational numbers in real-world situations. **(MC, GR)**

3 Marty and Jim ate an entire pizza. Jim ate 3 times as many pieces as Marty did. Which of the following represents the amount of pizza Jim ate?

A.

B.

C.

D.

Go On ▶

Analysis:

Choice B is correct. Since Jim ate 3 times as many pieces as Marty did, this means Jim ate 3/4 of the pizza and Marty only ate 1/4 of the pizza:

$$\left(\frac{1}{4} \times \frac{3}{1} = \frac{3}{4} \text{ and } \frac{1}{4} + \frac{3}{4} = \frac{3}{4} \text{ or 1 whole pizza} \right).$$

Answer B shows 3/4 of a pizza. Another way to look at this problem is to consider the number of pieces in the pizza. This pizza is cut into eighths. The boys ate all of the pieces, so consider the possible combinations. If Marty eats 1 piece, then Jim must eat 7 pieces. This can't be the right combination, since the problem states that Jim ate 3 times as much as Marty, not 7 times as much. If Marty eats 2 pieces, then Jim must eat 6 pieces. This is the correct combination since 3 x 2 pieces = 6 pieces and 6 + 2 = 8 pieces or one whole pizza. Choice B shows 6 pieces of pizza, the amount Jim ate.

Question **4** *assesses:*

Strand A: Number Sense, Concepts, and Operations

Standard 1: The student understands the different ways numbers are represented and used in the real world.

MA.A.1.3.4 Understands that numbers can be represented in a variety of equivalent forms, including integers, fractions, decimals, percents, scientific notation, exponents, radicals, and absolute value. **(MC, GR)**

Teaching Tip

Develop a grid with 4 to 9 rows and 5 or more columns. Label the first row "number", then choose at least three of the following number formats to label your other rows: integer, fraction, decimal, percent, scientific notation, exponent, radical, absolute value. Fill a variety of numbers into each column in the number row. Students may work in small groups to try to fill in each cell by rewriting the number in its appropriate form according to the row label (e.g., as an integer, as a fraction, as a decimal, etc.). The winning team is the one with the most cells filled in. It's OK if some formats are marked N.A. for Not Applicable or Too Hard. See example below.

number	4	$-\dfrac{7}{8}$			
closest integer	4	-1			
fraction	$\dfrac{8}{2}$	$-\dfrac{7}{8}$ or $-\dfrac{14}{16}$			
decimal	4.0	0.875			
per cent	400%	87.5%			
scientific notation	4.0×10^{0} or N.A.	8.75×10^{-1}			
exponent	2^{2}	Too Hard			
radical	$\sqrt{16}$	Too Hard			
absolute value	$\lvert 4 \rvert$ or $\lvert -4 \rvert$	$\left\lvert \dfrac{7}{8} \right\rvert$ or $\left\lvert -\dfrac{7}{8} \right\rvert$			

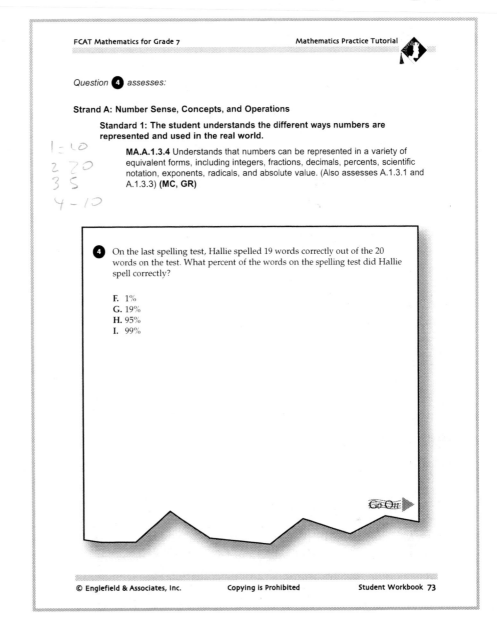

Question **4** *assesses:*

Strand A: Number Sense, Concepts, and Operations

> **Standard 1: The student understands the different ways numbers are represented and used in the real world.**
>
> > **MA.A.1.3.4** Understands that numbers can be represented in a variety of equivalent forms, including integers, fractions, decimals, percents, scientific notation, exponents, radicals, and absolute value. (Also assesses A.1.3.1 and A.1.3.3) **(MC, GR)**

4 On the last spelling test, Hallie spelled 19 words correctly out of the 20 words on the test. What percent of the words on the spelling test did Hallie spell correctly?

F. 1%
G. 19%
H. 95%
I. 99%

Go On ▶

Analysis:

Choice H is correct. Hallie spelled 19 words correctly out of 20 possible, so 19/20 represents the fraction of words spelled correctly. To convert the fraction to a percent, divide its numerator by its denominator and multiply by 100:

$$19 \div 20 = 0.95; \; 0.95 \times 100 = 95\%.$$

All the other choices are incorrect because they are miscalculations or misunderstandings of percents.

Question **5** *assesses:*

Strand A: Number Sense, Concepts, and Operations

Standard 2: The student understands number systems.

MA.A.2.3.1 Understands and uses exponential and scientific notation.**(MC, GR)**

Teaching Tips

Make sure that students understand these facts and the effects of squaring different kinds of numbers:

1. Squaring a number means multiplying that number by itself.

2. Squaring any number, whether positive or negative, always results in a positive number. For example, $7^2 = 49$ and $(-7)^2 = 49$.

3. Squaring any fraction between 0 and 1 always results in a smaller number. First square the numerator, then square the denominator.
For example, $\left(\dfrac{1}{4}\right)^2 = \dfrac{1}{4} \times \dfrac{1}{4} = \dfrac{1}{16}$.

4. A negative exponent does not mean a negative number, it means a reciprocal. So, a number raised to the negative 2 power means the reciprocal of the number squared.
For example, $5^{-2} = \dfrac{1}{5^2} = \dfrac{1}{25}$.

• Review scientific notation with students. Scientific notation is a compact way of writing very large or very small numbers. It is usually used in various branches of science. A number written in scientific notation has two parts: a decimal fraction between 1 and 10 and some power of 10. Students usually know that when you convert a number in standard notation to scientific notation, you must move the decimal point the same number of spaces as the power of 10. However, many students make mistakes in the direction they move the decimal. They also often think that a negative exponent means a negative number when it really means a fraction.

Sample explanation of scientific notation: If scientific notation confuses you, use this method for deciding which way to move the decimal. First notice were the decimal place is. All numbers have a decimal place even if it's not actually printed in them. Move the decimal place to the right of the first significant digit. The first significant digit is the first non-zero digit you come to reading from left to right.

For example: In 778,330,000, the first significant digit is 7 so I want to move my decimal point from the end of the number to between the sevens. The decimal moves 8 places, so the exponent should be 8.

decimal point is here

778,330,000
8 7 6 5 4 3 2 1

need decimal point to be here
(since 7 is the first significant digit)

If the number you're converting looks bigger than one, use a positive exponent. If it looks smaller than one, use a negative exponent. Consider this example: 778,330,000. Is this number bigger than one or less than one? Bigger than one, right? So the 8 in the exponent should be positive

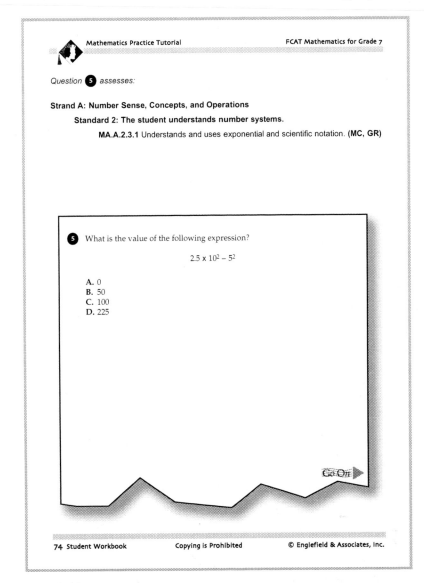

Mathematics Practice Tutorial   FCAT Mathematics for Grade 7

Question **5** *assesses:*

Strand A: Number Sense, Concepts, and Operations

    Standard 2: The student understands number systems.

        MA.A.2.3.1 Understands and uses exponential and scientific notation. **(MC, GR)**

5 What is the value of the following expression?

$$2.5 \times 10^2 - 5^2$$

A. 0
B. 50
C. 100
D. 225

Go On

74 Student Workbook   Copying is Prohibited   © Englefield & Associates, Inc.

Analysis:

Choice D is correct. This is an example of a power or exponential expression. Exponents have the second highest priority in order of operations, just after parentheses. The 10 in the expression 10^2 is called the base, while the 2 is called the power or exponent. The power tells how many times the base should be multiplied by itself. Since this power is two, two factors of 10 must be multiplied ($10^2 = 10 \times 10 = 100$). Similarly, $5^2 = 5 \times 5 = 25$. This is also called squaring a number, so ten squared is one hundred, and five squared is twenty-five. Use PEMDAS to help you remember the correct order of operations. P is for parentheses, but there are none. E is for exponents so they are done first:

$$2.5 \times 10^2 - 5^2 = 2.5 \times 100 - 25.$$

M is for multiplication, so it is done next:

$$2.5 \times 100 - 25 = 250 - 25.$$

The only operation left is the subtraction:

$$250 - 25 = 225.$$

Note: A common mistake many students make is to misinterpret the meaning of the exponent. Remember, 10^2 does not mean to multiply 10 times 2, it means to multiply 10 by itself: $10 \times 10 = 100$. The other choices are miscalculations and are incorrect.

Question **6** *assesses:*

Strand A: Number Sense, Concepts, and Operations

Standard 3: The student understands the effects of operations on numbers and the relationships among these operations, selects appropriate operations, and computes for problem solving.

MA.A.3.3.1 Understands and explains the effects of addition, subtraction, multiplication, and division on whole numbers and fractions, including mixed numbers and decimals, including the inverse relationships of positive and negative numbers. **(MC)**

Teaching Tip

Students often have difficulty remembering whether multiplying or dividing signed numbers produces a positive or a negative result. As a whole class activity, prepare 6 large flash cards, two marked with a minus (–) symbol, two marked with a plus (+) symbol, one card with a multiplication (x) symbol, and the final card with a division symbol (÷). Test students on their knowledge of dividing integers by placing the card with the division sign on the blackboard ledge. Then place two negative symbols (representing the dividend and the divisor) on each side. Students should respond that if the dividend and the divisor are negative, the quotient is positive. Next, place negative and positive symbols on each side of the division symbol. Students should respond that if either the dividend or the divisor is negative, the quotient is negative. Explain to students if the signs are the same, the quotient is positive. If the signs are different, the quotient is negative. After the students have mastered division of negative and positive integers, repeat this skill drill to review multiplying negative and positive integers. Don't forget to use fractions, decimals, whole numbers, and mixed numbers to further explain this process. As a follow-up activity, prepare a worksheet or blackboard activity with 20 or so mixed multiplication and division problems. Include a wide variety of both positive and negative number formats. Ask students to indicate as quickly as possible whether the result of the given operation is positive or negative without actually solving the problem.

Question **6** *assesses:*

Strand A: Number Sense, Concepts, and Operations

> **Standard 3: The student understands the effects of operations on numbers and the relationships among these operations, selects appropriate operations, and computes for problem solving.**

>> **MA.A.3.3.1** Understands and explains the effects of addition, subtraction, multiplication, and division on whole numbers and fractions, including mixed numbers and decimals, including the inverse relationships of positive and negative numbers. **(MC)**

6 In the expression below, using what operation will result in the **greatest** number?

$$\frac{5}{12} \ \square \ \frac{5}{6}$$

F. subtraction
G. addition
H. division
I. multiplication

Go On ▶

Analysis:

Choice G is correct. When these fractions are added together, the resulting number is greater than one:

$$\frac{5}{12} + \frac{5}{6} = \frac{5}{12} + \frac{5}{6} \cdot \frac{2}{2} = \frac{5}{12} + \frac{10}{12} = \frac{15}{12}$$

The other three operations result in a number less than one. Choice F is incorrect because the result of subtraction is $-\frac{5}{12}$:

$$\frac{5}{12} - \frac{5}{6} = \frac{5}{12} - \frac{5}{6} \cdot \frac{2}{2} = \frac{5}{12} - \frac{10}{12} = -\frac{5}{12}$$

Choice H is incorrect because the result of division is $\frac{1}{2}$:

$$\frac{5}{12} \div \frac{5}{6} = \frac{5}{12} \times \frac{6}{5} = \frac{30}{60} = \frac{1}{2}$$

Choice I is incorrect because the result of multiplication is $\frac{25}{72}$:

$$\frac{5}{12} \times \frac{5}{6} = \frac{25}{72}$$

Question **7** *assesses:*

Strand A: Number Sense, Concepts, and Operations

Standard 3: The student understands the effects of operations on numbers and the relationships among these operations, selects appropriate operations, and computes for problem solving.

MA.A.3.3.2 Selects the appropriate operation to solve problems involving addition, subtraction, multiplication, and division of rational numbers, ratios, proportions, and percents, including the appropriate application of the algebraic order of operations. **(MC, GR)**

Teaching Tips

- Students may be great at addition, subtraction, multiplication, and division, yet have difficulty simplifying mathematical expressions using the order of operations. Teach students to remember the conventional order of operations by memorizing the phrase, "**P**lease **E**xcuse **M**y **D**ear **A**unt **S**ally." The letters in bold form the word PEMDAS, each letter of which stands for a different operation in the exact order of how they should be computed: operations in Parentheses have the highest priority; numbers raised to Exponents have the second highest priority; Multiplication and Division both have the third highest priority; Addition and Subtraction both have the lowest priority. If there are two operations with the same priority in an expression, they should be worked in order from left to right.

Example:

$10 \div (6 + 4) \times 3^3 - 5$	Add inside Parentheses to get . . .
$10 \div 10 \times 3^3 - 5$	Raise numbers to Exponents to get . . .
$10 \div 10 \times 27 - 5$	Multiply and Divide from left to right to get . . .
$27 - 5$	Subtract to get . . .
22	

- Remind students that many calculators do not automatically use the correct order of operations. They may require using complex sets of nested parentheses. Show several examples of expressions with nested parentheses and how to correctly simplify them. Demonstrate how expressions simplified with the wrong order of operations produce incorrect answers.

- Remind students that sets of brackets [ ] as well as fraction bars count as parentheses. For example:

$\dfrac{9 - 2 \times 3}{3^3 - 6}$ is treated as $\dfrac{(9 - 2 \times 3)}{(3^3 - 6)}$, even though the numerator and denominator are not actually enclosed in parentheses. This means that the numerator and denominator are treated as separate expressions until the very last operation, the division implied by the fraction bar. So, the expression

$\dfrac{9 - 2 \times 3}{3^3 - 6}$ simplifies to $\dfrac{3}{21}$ or $\dfrac{1}{7}$.

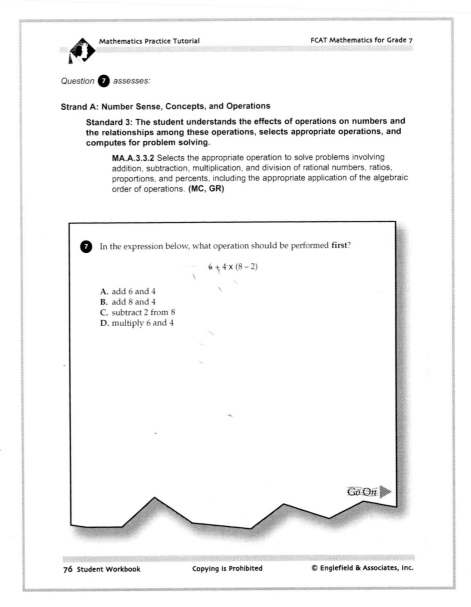

Mathematics Practice Tutorial   FCAT Mathematics for Grade 7

Question **7** *assesses:*

Strand A: Number Sense, Concepts, and Operations

> **Standard 3: The student understands the effects of operations on numbers and the relationships among these operations, selects appropriate operations, and computes for problem solving.**
>
> > **MA.A.3.3.2** Selects the appropriate operation to solve problems involving addition, subtraction, multiplication, and division of rational numbers, ratios, proportions, and percents, including the appropriate application of the algebraic order of operations. **(MC, GR)**

7 In the expression below, what operation should be performed **first**?

$$6 + 4 \times (8 - 2)$$

A. add 6 and 4
B. add 8 and 4
C. subtract 2 from 8
D. multiply 6 and 4

Go On

Analysis:

Choice C is correct. The proper order of operations requires that any operation within a set of parentheses should be performed first. If there is more than one set of parentheses, work "nested" parentheses (one set of parentheses inside another set of parentheses) "from the inside out". If the sets of parentheses are not nested, work them from left to right. In this case, the subtraction inside the parentheses is done first, then the result is multiplied by 4, and finally the product is added to the 6:

$$6 + 4 \times (8 - 2) = 6 + 4 \times 6 = 6 + 24 = 30.$$

The other choices are miscalculations and are incorrect.

Question **8** *assesses:*

Strand A: Number Sense, Concepts, and Operations

Standard 3: The student understands the effects of operations on numbers and the relationships among these operations, selects appropriate operations, and computes for problem solving.

MA.A.3.3.3 Adds, subtracts, multiplies, and divides whole numbers, decimals, and fractions, including mixed numbers, to solve real-world problems, using appropriate methods of computing, such as mental mathematics, paper and pencil, and calculator. **(MC, GR)**

Teaching Tip

Review how to convert between decimals and percents and show students this pneumonic device to help them remember the process.

Explanation: Convert percents to decimals by dividing by 100. This moves the decimal two places to the left. Convert decimals to percents by multiplying by 100. This moves the decimal two places to the right. If you get confused changing decimals to percents or percents to decimals, it is useful to use this little device. Write down a D for decimal and P for percent. Always write the letter D first since D comes before P in the alphabet. If you want to convert decimals to percents, draw an arrow from the D to the P as so:

$$D \longrightarrow P$$

If you want to convert percents to decimals, draw an arrow from the P to the D as so:

$$D \longleftarrow P$$

When converting from decimals to percents or from percents to decimals you always move the decimal point two places. These two diagrams tell you which way to move it.

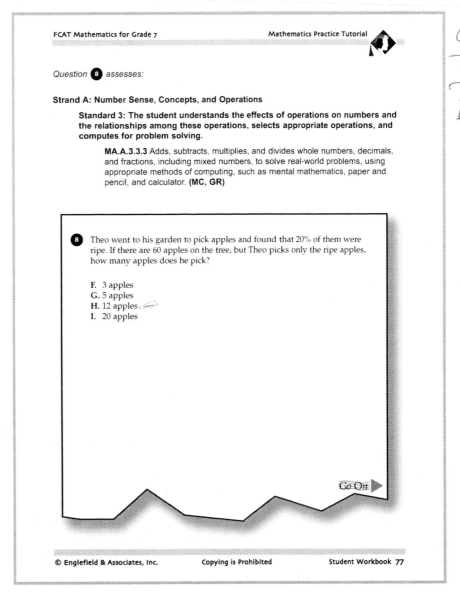

Question **8** *assesses:*

Strand A: Number Sense, Concepts, and Operations

Standard 3: The student understands the effects of operations on numbers and the relationships among these operations, selects appropriate operations, and computes for problem solving.

MA.A.3.3.3 Adds, subtracts, multiplies, and divides whole numbers, decimals, and fractions, including mixed numbers, to solve real-world problems, using appropriate methods of computing, such as mental mathematics, paper and pencil, and calculator. **(MC, GR)**

8 Theo went to his garden to pick apples and found that 20% of them were ripe. If there are 60 apples on the tree, but Theo picks only the ripe apples, how many apples does he pick?

F. 3 apples
G. 5 apples
H. 12 apples
I.   20 apples

Go On ▶

Analysis:

Choice H is correct. Since Theo picks only the ripe apples, he picks 20% of the 60 on the tree. This is represented by the expression: 20% x 60. Simplify this expression by converting the percent to its decimal equivalent and multiplying: 20% x 60 = 0.20 x 60 = 12 apples. Alternatively, you could realize that taking 20% of a number is the same as finding one fifth of it because 20% is equal to the fraction 1/5. To find one fifth of 60, divide 60 by 5: 60 apples ÷ 5 = 12 apples. Choice F is incorrect because 3 is just 5% of 60, not 20% of 60. Choice G is incorrect because 5 is a little more than 8% of 60, not 20% of 60. Choice I is incorrect because 20 is about 33% of 60, not 20% of 60.

Question **9** *assesses:*

Strand A: Number Sense, Concepts, and Operations

Standard 4: The student uses estimation in problem solving and computation.

MA.A.4.3.1 Uses estimation strategies to predict results and to check the reasonableness of results. (Also assesses A.4.2.1, B.2.3.1, and B.3.3.1) **(MC)**

Teaching Tips

- Remind students that they may lose points on the FCAT if they provide an exact answer to a question that calls for an estimate. Review rounding with students. Emphasize that estimation usually requires rounding which makes it easier to work with numbers. When rounding, check the digit to the right of the number required for the proper level of precision. If the digit is 5 or greater, round up, otherwise round down. For example, 2,173 rounded to the nearest hundred is 2,200 because the digit to the right of the hundreds place is greater than 5, so the number in the hundreds place gets rounded up. Give students practice rounding the same number to different levels of precision and discuss why different levels of precision may be needed. For example, round 9,357 to the nearest ten, to the nearest hundred, and to the nearest thousand.

- Discuss with students the many reasons for estimating instead of calculating an exact answer. Among these are:

    1. You may not need an exact answer. You may only need to fix a value in a rough or general way.

    2. You may not want to invest the time, effort, or expense necessary to calculate an exact answer.

    3. It may be impossible to calculate an exact answer because the data you have is incomplete or imperfect.

    4. You may need to use some sort of sampling or polling technique because the data you are trying to measure keeps changing. In the time it would take to get an exact answer, the data changes so much that conclusions drawn from it are useless.

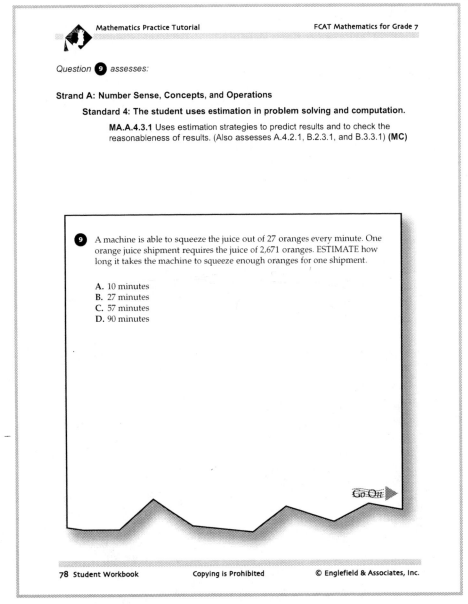

Mathematics Practice Tutorial   FCAT Mathematics for Grade 7

Question **9** *assesses:*

Strand A: Number Sense, Concepts, and Operations

    Standard 4: The student uses estimation in problem solving and computation.

        MA.A.4.3.1 Uses estimation strategies to predict results and to check the reasonableness of results. (Also assesses A.4.2.1, B.2.3.1, and B.3.3.1) **(MC)**

9   A machine is able to squeeze the juice out of 27 oranges every minute. One orange juice shipment requires the juice of 2,671 oranges. ESTIMATE how long it takes the machine to squeeze enough oranges for one shipment.

    A. 10 minutes
    B. 27 minutes
    C. 57 minutes
    D. 90 minutes

Go On ▶

78 Student Workbook   Copying is Prohibited   © Englefield & Associates, Inc.

Analysis:

Choice D is correct. It takes about 2,700 oranges for a shipment of orange juice and the juicer can squeeze about 30 oranges a minute. To find the time it takes the juicer to squeeze enough oranges for a shipment, divide the number of oranges in a shipment by the juicer's rate per minute: 2,700 oranges ÷ 30 oranges per minute ≈ 90 minutes to squeeze 2,700 oranges. The other choices are miscalculations and are incorrect.

Question **10** assesses:

Strand A: Number Sense, Concepts, and Operations

Standard 4: The student uses estimation in problem solving and computation.

MA.A.5.3.1 Uses concepts about numbers, including primes, factors, and multiples, to build number sequences. **(MC, GR)**

Teaching Tip

• Remind students that all whole composite numbers can be uniquely written as the product of only prime numbers. For example, the expression $2 \times 3^3 \times 7^2$ represents the number 2,646 as the product of primes.

• Review how to find the prime factors of a number, as well as how to find the Greatest Common Factor (GCF) of two numbers using prime divisors. Show students how to set up a tree diagram by attempting to divide each number by successively larger primes. For example, try dividing 90 by 2 to get $2 \times 45 = 90$. Since 45 cannot be divided by 2, try dividing it by 3 to get $3 \times 15 = 45$, so $2 \times 3 \times 15 = 90$. Now try dividing 15 by 3 to get $3 \times 5 = 15$, so $2 \times 3 \times 3 \times 5 = 90$ or $2 \times 3^2 \times 5 = 90$. Repeat this process with 210. See the tree diagrams below which show 90 and 210 factored down to primes.

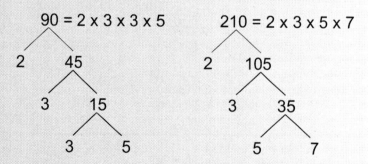

Now that both numbers are factored down to primes, find the product of all the factors that they have in common. For example, the diagram shows that both 90 and 210 have one 2, one 3, and one 5. These are the factors that they have in common, therefore the Greatest Common Factor of these two numbers is $2 \times 3 \times 5 = 30$. Have students take turns finding all the prime factors of a number and the GCF of two or more numbers using this method on an overhead projector or on the blackboard.

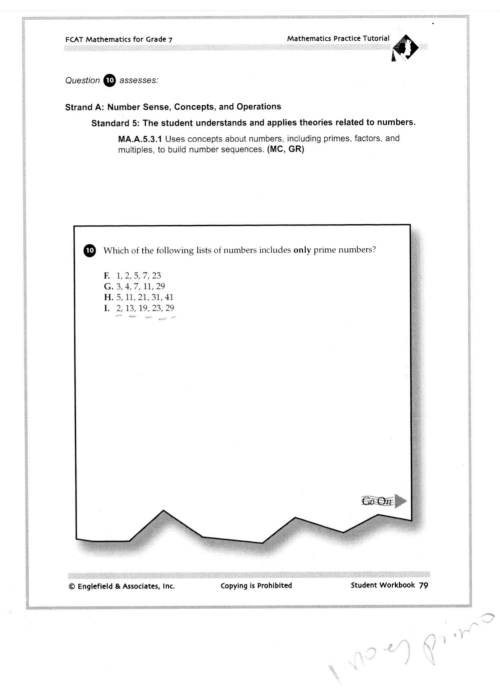

Question **10** *assesses:*

Strand A: Number Sense, Concepts, and Operations

   Standard 5: The student understands and applies theories related to numbers.

      MA.A.5.3.1 Uses concepts about numbers, including primes, factors, and multiples, to build number sequences. **(MC, GR)**

10 Which of the following lists of numbers includes **only** prime numbers?

   F. 1, 2, 5, 7, 23
   G. 3, 4, 7, 11, 29
   H. 5, 11, 21, 31, 41
   I. 2, 13, 19, 23, 29

Go On ▶

Analysis:

Choice I is correct. A prime number is any number (except 1) whose only factors are itself and one. The only divisors of the numbers in Choice I are themselves and 1, so they are all prime. Choice F is incorrect because 1 is not a prime number. Choice G is incorrect because 4 is not prime since it is divisible by 2. Choice H is incorrect because 21 is not prime since it is divisible by 3 and 7.

Question **11** *assesses:*

Strand B: Measurement

Standard 1: The student measures quantities in the real world and uses the measures to solve problems.

MA.B.1.3.1 Uses concrete and graphic models to derive formulas for finding perimeter, area, surface area, circumference, and volume of two- and three-dimensional shapes, including rectangular solids and cylinders. (Also assesses B.1.2.2 and B.2.3.1) **(MC, GR)**

Teaching Tip

- Set out a variety of objects on a table such as a pop can, a book, a box, a cone-shaped disposable cup, and a pyramid. Demonstrate how to sketch the net of an object. (An object's net is an arrangement of shapes on a single sheet of paper that are joined at the edges and can be cut out and folded to become the faces of the object. See the sketch of a cylinder's net below.)

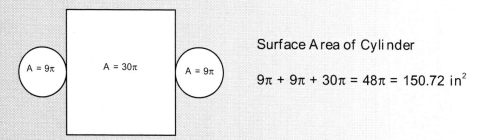

Surface Area of Cylinder

$9\pi + 9\pi + 30\pi = 48\pi = 150.72 \text{ in}^2$

Divide the class into groups, each with a pencil, paper, straightedge, and measuring tape. Have students measure the dimensions of each face on each object and use their measurements to sketch a net of each object. Remind students that the face edge lengths should be labeled on their nets. Ask groups to calculate both the surface area and the volume of each object. After the students have completed their measurements, drawings, and calculations, have each group present its findings.

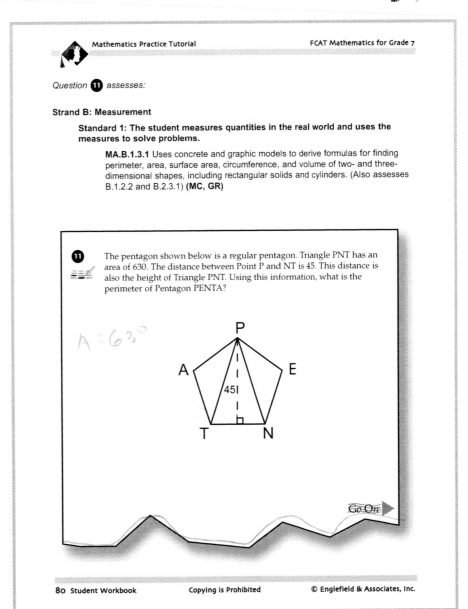

Question **11** *assesses:*

Strand B: Measurement

Standard 1: The student measures quantities in the real world and uses the measures to solve problems.

MA.B.1.3.1 Uses concrete and graphic models to derive formulas for finding perimeter, area, surface area, circumference, and volume of two- and three-dimensional shapes, including rectangular solids and cylinders. (Also assesses B.1.2.2 and B.2.3.1) **(MC, GR)**

11 The pentagon shown below is a regular pentagon. Triangle PNT has an area of 630. The distance between Point P and NT is 45. This distance is also the height of Triangle PNT. Using this information, what is the perimeter of Pentagon PENTA?

Go On ▶

80 Student Workbook   Copying is Prohibited   © Englefield & Associates, Inc.

Analysis:

The correct answer is 140. A regular pentagon has 5 congruent sides, so the perimeter of this pentagon is simply 5 times the length of one of its sides. The area and height of Triangle PNT is given, so this information and the formula for the area of a triangle can be used to find its base which is also the length of the pentagon's side:

$$A = \frac{1}{2}bh \text{ where } A = 630 \text{ and } h = 45; \; 630 = \frac{1}{2} \times b \times 45; \; 630 = 22.5 \times b; \; b = 28.$$

The perimeter of the pentagon is five times this length: 5 x 28 = 140. The perimeter of the pentagon is 140.

Question **12** *assesses:*

Strand B: Measurement

> **Standard 1: The student measures quantities in the real world and uses the measures to solve problems.**
>
> > **MA.B.1.3.2** Uses concrete and graphic models to derive formulas for finding rates, distance, time, and angle measures. (Also assesses B.1.2.2 and B.2.3.1) **(MC)**

Teaching Tip

Have students use pretzel sticks to construct several regular polygons, identify various angles, and estimate the size of constructed angles. For example, place two sticks on a table to simulate vertical angles. Place four pieces of masking tape on the table under the angles and ask students to write the type of each angle on the tape and also an estimate of their sizes. Their responses should indicate that two of the angles are acute and two of the angles are obtuse, unless the sticks are perpendicular to one another and all of the angles are congruent right angles. Ask students to check their angle measurements with a protractor. Repeat this exercise several times during the year to help students develop the ability to estimate an angle's approximate size by sight. You can also ask students to identify complementary angles and supplementary angles using this method. At the completion of the lesson, students can snack on the leftover pretzel sticks.

Question **12** *assesses:*

Strand B: Measurement

Standard 1: The student measures quantities in the real world and uses the measures to solve problems.

MA.B.1.3.2 Uses concrete and graphic models to derive formulas for finding rates, distance, time, and angle measures. (Also assesses B.1.2.2 and B.2.3.1) **(MC)**

12 At which point in the figure below is an **obtuse** angle located?

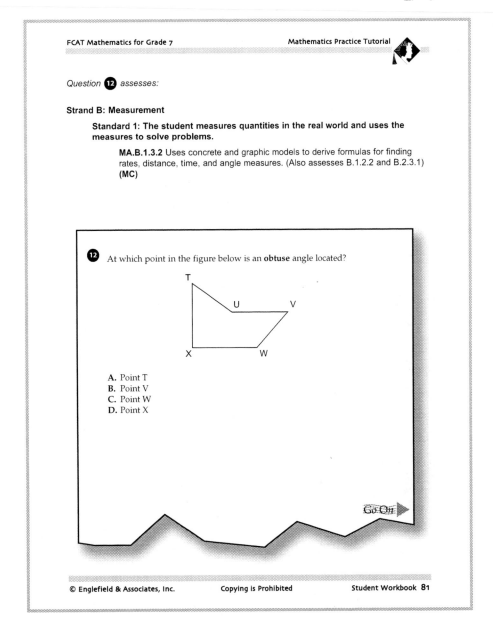

A. Point T
B. Point V
C. Point W
D. Point X

Go On ▶

Analysis:

Choice C is correct. An obtuse angle is an angle of more than 90° but less than 180°. Choices A and B are incorrect because the angles at Points T and V are acute angles which are angles less than 90° but greater than 0°. Choice D is incorrect because the angle formed at Point X is a right angle. Right angles measure exactly 90°.

Question **13** *assesses:*

Strand B: Measurement

> **Standard 1: The student measures quantities in the real world and uses the measures to solve problems.**
>
> > **MA.B.1.3.3** Understands and describes how the change of a figure in such dimensions as length, width, height, or radius affects its other measurements such as perimeter, area, surface area, and volume. (Also assesses C.2.3.1) **(MC, GR)**

Teaching Tips

- Items on the Grade 7 FCAT for Mathematics may require students to determine how an increase or decrease in the dimensions (e.g., length, width, height, radius) of an object affects its other measurements (e.g., perimeter, area, surface area, volume). In order to be successful in this benchmark, students should be able to compute length, perimeter, circumference, area, surface area, and volume of figures.

- To practice problems involving two-dimensional objects, give each student construction paper, tape, scissors, a pencil, a compass, and a ruler. Assign dimensions for circles, triangles, squares, and rectangles. Ask students to measure and cut out these objects. On the back of each object, have the students calculate the shape's perimeter and area. Next, have students either increase or decrease each object by a given amount (add strips of paper or cut away). Students then determine how the perimeter and area have changed. For three-dimensional objects, use wooden blocks and measure face perimeters, surface area, and volume.

36 inches

24 ins

36 × 24 = 464

2 ancho

3 cargo

2 inchs tuck

3

12 × 2 = 24

12 + 3 = 36

2

9:

4

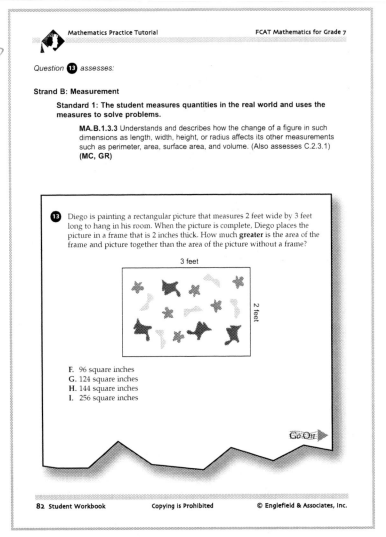

Mathematics Practice Tutorial   FCAT Mathematics for Grade 7

Question **13** *assesses:*

Strand B: Measurement

Standard 1: The student measures quantities in the real world and uses the measures to solve problems.

MA.B.1.3.3 Understands and describes how the change of a figure in such dimensions as length, width, height, or radius affects its other measurements such as perimeter, area, surface area, and volume. (Also assesses C.2.3.1) **(MC, GR)**

13 Diego is painting a rectangular picture that measures 2 feet wide by 3 feet long to hang in his room. When the picture is complete, Diego places the picture in a frame that is 2 inches thick. How much **greater** is the area of the frame and picture together than the area of the picture without a frame?

3 feet

2 feet

F. 96 square inches
G. 124 square inches
H. 144 square inches
I.   256 square inches

Go On ▶

82 Student Workbook   Copying is Prohibited   © Englefield & Associates, Inc.

Analysis:

Choice I is correct. Begin by converting the dimensions of the picture from feet to inches by multiplying both the length and the width by 12:

12 x 2 feet = 24 inches; 12 x 3 feet = 36 inches.

Since the picture is 24 inches wide and 36 inches long, its area is 864 square inches:

24 inches x 36 inches = 864 square inches.

When the thickness of the frame is added to the dimensions of the picture, 2 inches of length is added to each of the picture's four sides. This makes both the length and width of the picture increase by 4 inches when the frame is placed around it. The new dimensions are 28 inches wide by 40 inches long for an area of 1,120 square inches:

28 inches x 40 inches = 1,120 square inches.

The difference between the area of the frame and the picture together and the picture alone is 256 square inches:

1,120 square inches – 864 square inches = 256 square inches.

The other choices are miscalculations and are incorrect.

Question **14** *assesses:*

Strand B: Measurement

Standard 1: The student measures quantities in the real world and uses the measures to solve problems.

MA.B.1.3.4 Constructs, interprets, and uses scale drawings such as those based on number lines and maps to solve real-world problems. (Also assesses B.2.3.1) **(MC, GR)**

Teaching Tips

- Scale drawings can be used to solve real-world problems, including distance. Measurement may be in either metric or customary units. Provide students with a variety of maps. Have them plot distances from one designated point to another. Using the scale, have them calculate the distance from one point to another. Ask students to check their calculations using one of the Internet mapping sites such as www.mapblast.com, www.mapquest.com, or maps.yahoo.com. Discuss the difference between road distance versus distance "as the crow files."

- Divide students into groups. Assign each group the task of creating a scale drawing of their classroom and computing various measurements such as perimeter and area of the room, or distance from one point of the room to another. Be sure they include a scale to indicate distance. Ask groups to present and compare their drawings and calculations.

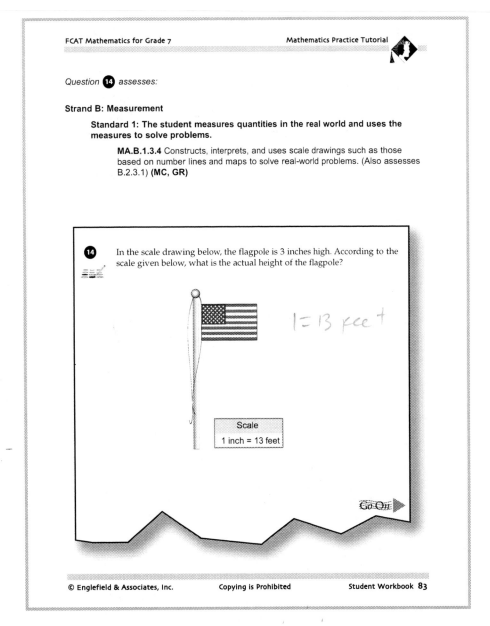

Analysis:

The correct answer is 39 feet. The flagpole shown is 3 inches high. According to the scale, each inch in the picture represents 13 feet of actual length. Since there are 3 inches, multiplying 3 by 13 will give the actual height of 39 feet:

$$3 \times 13 \text{ feet} = 39 \text{ feet.}$$

Question **15** *assesses:*

Strand B: Measurement

Standard 2: The student compares, contrasts, and converts within systems of measurement (both standard/nonstandard and metric/customary).

MA.B.2.3.1 Uses direct (measured) and indirect (not measured) measures to compare a given characteristic in either metric or customary units. **(MC, GR)**

Teaching Tip

It's important for students to get an intuitive understanding about the relative sizes of linear measure units, area units, and volume units and how these three types relate to each other. The most important standard linear units are: inches, feet, yards, and miles. The most important metric linear units are: millimeters, centimeters, meters, and kilometers. Give students lots of opportunities to use these measures in normal classwork. A group scavenger hunt is also a good way to help build this intuition. Put students into small groups of three or four with a selection of old magazines. Have each group find at least one picture of an object that can be appropriately measured with each of the eight units above. Each group should then prepare a poster to share their findings with the whole class. Discuss whether each picture-unit pair is appropriate and whether or not another picture-unit pair would serve just as well. To begin with, this activity should focus exclusively on linear measure until students become very familiar with the units. Later, similar lessons can be attempted with weight, area, and volume units.

Question **15** *assesses:*

Strand B: Measurement

> **Standard 2: The student compares, contrasts, and converts within systems of measurement (both standard/nonstandard and metric/customary).**
>
> > **MA.B.2.3.1** Uses direct (measured) and indirect (not measured) measures to compare a given characteristic in either metric or customary units. **(MC, GR)**

15 Jimmy and his friends are having a contest to see how far they can hit a baseball. Jimmy was able to hit his ball 3 times as far as Gwen. Art only managed to hit his ball half as far as Kelly. Gwen's ball landed just past Kelly's ball. Which is the correct order of persons who hit the ball from farthest to shortest distance?

A. Jimmy, Gwen, Art, Kelly
B. Art, Kelly, Gwen, Jimmy
C. Jimmy, Gwen, Kelly, Art
D. Jimmy, Art, Gwen, Kelly

Go On ▶

Analysis:

Choice C is correct. The first information given tells us that Jimmy hit his ball farther than Gwen's. The next statement tells us that Kelly's ball went further than Art's. Finally, we are told that Gwen's ball went further than Kelly's. So, Jimmy hit his ball farther than Gwen, who hit her ball farther than Kelly, who hit her ball farther than Art. All the other choices result from an incorrect interpretation of the data and are incorrect.

Question **16** *assesses:*

Strand B: Measurement

Standard 2: The student compares, contrasts, and converts within systems of measurement (both standard/nonstandard and metric/customary).

MA.B.2.3.2 Solves problems involving units of measure and converts answers to a larger or smaller unit within either the metric or customary system.
(MC, GR)

Teaching Tips

• Students should be able to solve problems by converting between units within the same measurement system (e.g., metric or customary). Have students make equivalency posters for length and weight units. Place finished posters in a prominent area of the classroom. On an overhead transparency, give students conversion problems to solve (e.g., converting ounces to pounds; miles to yards; kiloliters to liters). Ask students to come to the overhead and have them write out their solutions to these problems.

• Although students may know how to convert between two units in the same measurement system, they may not be as comfortable with expressions involving mixed units. Have students practice solving these types of expressions. For example: Jan walked 94 yards, 67 feet, and 14 inches in 2 minutes and 9 seconds. Alex walked 87 yards, 94 feet, and 75 inches in 1 minute and 174 seconds. Who traveled a greater distance? What was the difference in the two lengths walked? Who walked for a greater amount of time?

1 yard = 3 feet
1 feet = 12 inches

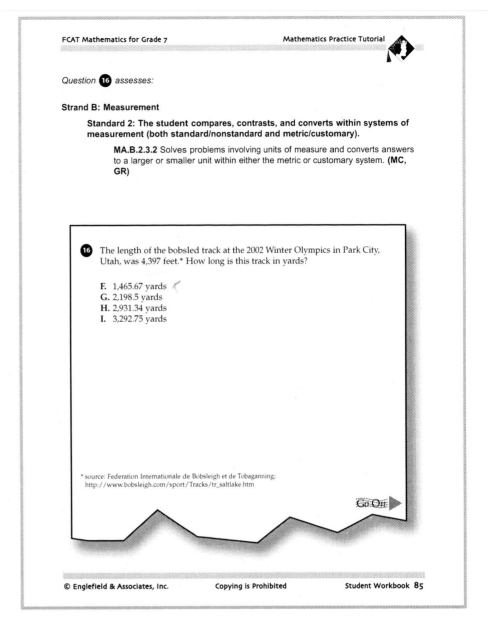

Question **16** *assesses:*

Strand B: Measurement

> **Standard 2: The student compares, contrasts, and converts within systems of measurement (both standard/nonstandard and metric/customary).**
>
> > **MA.B.2.3.2** Solves problems involving units of measure and converts answers to a larger or smaller unit within either the metric or customary system. **(MC, GR)**

16 The length of the bobsled track at the 2002 Winter Olympics in Park City, Utah, was 4,397 feet.* How long is this track in yards?

F. 1,465.67 yards
G. 2,198.5 yards
H. 2,931.34 yards
I. 3,292.75 yards

* source: Federation Internationale de Bobsleigh et de Tobaganning; http://www.bobsleigh.com/sport/Tracks/tr_saltlake.htm

Go On ▶

Analysis:

Choice F is correct. One yard is equal to 3 feet. To convert feet to yards, divide the number of feet by 3:

$$4,397 \div 3 = 1,465.67 \text{ yards.}$$

The other choices are miscalculations and are incorrect.

Question **17** *assesses:*

Strand B: Measurement

Standard 3: The student estimates measurements in real-world problem situations.

MA.B.3.3.1 Solves real-world and mathematical problems involving estimates of measurements including length, time, weight/mass, temperature, money, perimeter, area, and volume, in either customary or metric units. **(MC)**

Teaching Tip

Give groups of students specific building dimensions and a materials list for a project such as an outdoor deck, a doghouse, or a garden shed. Have them research the cost of the items on the materials list and estimate cost of building the project. Ask students to calculate the perimeter and area of the project as well as the estimated cost per square foot, etc. Allow students to suggest ways to research item cost. Some may choose to call a building supply store while others may attempt to complete their research on the Web. Compare each group's calculations and estimated costs.

© Englefield & Associates, Inc.

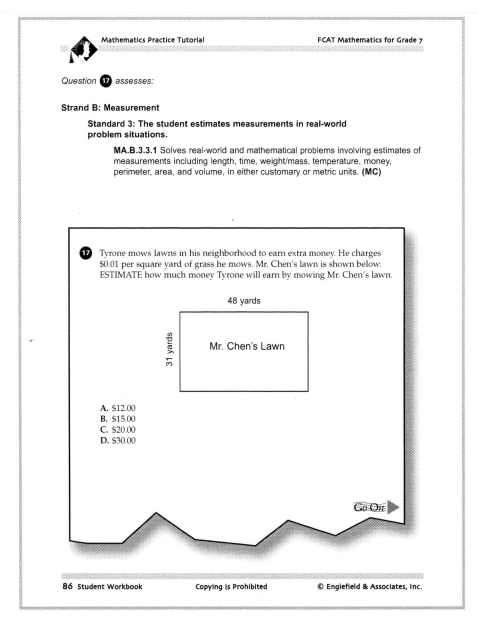

Question **17** *assesses:*

Strand B: Measurement

Standard 3: The student estimates measurements in real-world problem situations.

MA.B.3.3.1 Solves real-world and mathematical problems involving estimates of measurements including length, time, weight/mass, temperature, money, perimeter, area, and volume, in either customary or metric units. **(MC)**

17 Tyrone mows lawns in his neighborhood to earn extra money. He charges $0.01 per square yard of grass he mows. Mr. Chen's lawn is shown below. ESTIMATE how much money Tyrone will earn by mowing Mr. Chen's lawn.

48 yards

31 yards

Mr. Chen's Lawn

A. $12.00
B. $15.00
C. $20.00
D. $30.00

86 Student Workbook   Copying is Prohibited   © Englefield & Associates, Inc.

Analysis:

Choice B is correct. The lawn is approximately 30 yards by 50 yards or 1,500 square yards:

30 yards x 50 yards = 1,500 square yards.

Multiply Tyrone's price of $0.01 by the total number of square yards:

1,500 x $0.01 = $15.00.

The other choices are either guesses or miscalculations and are incorrect.

Question **18** *assesses:*

Strand C: Geometry and Spatial Sense

Standard 1: The student describes, draws, identifies, and analyzes two- and three-dimensional shapes.

MA.C.1.3.1 Understands the basic properties of, and relationships pertaining to, regular and irregular geometric shapes in two and three dimensions. (Also assesses C.1.2.1) **(MC)**

Teaching Tip

Give students construction paper, scissors, rulers, and protractors. Ask students to create rectangles and squares with assigned lengths and widths. Have students measure each of the shape's interior angles. Have them also label and find the sum of the interior angles of each shape. They should find that the sum of the four angles (90° each) is 360° for each figure. Review that regardless of the differences in their length and width, squares and rectangles have internal angle sums of 360°. Next, have students create any size quadrilateral that does not incorporate right angles. Ask them to measure each angle and compute the sum of the interior angles. By folding the quadrilateral on a diagonal, two triangles are formed. The sum of the interior angles of any triangle is 180°, so the sum of the 2 triangles is 360°. Have students verify their findings with a protractor.

Question **18** *assesses:*

Strand C: Geometry and Spatial Sense

> **Standard 1: The student describes, draws, identifies, and analyzes two- and three-dimensional shapes.**
>
> > **MA.C.1.3.1** Understands the basic properties of, and relationships pertaining to, regular and irregular geometric shapes in two and three dimensions. (Also assesses C.1.2.1) **(MC)**

18 In Quadrilateral ABCE below, Angle ABC is divided into two congruent angles by BD. Angle BCE is 60°. What is the measure of Angle ABD?

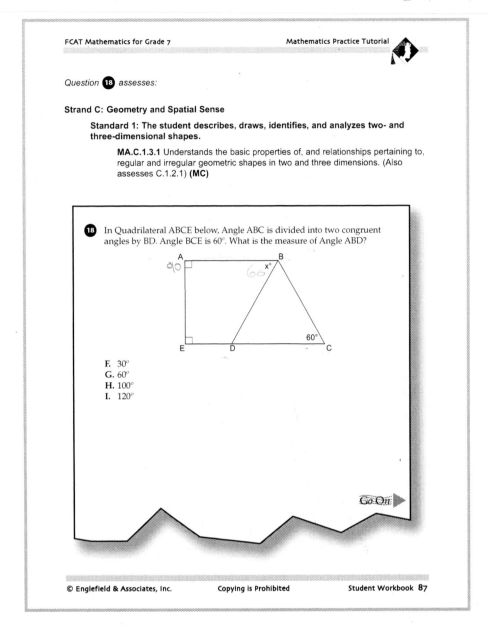

F. 30°
G. 60°
H. 100°
I. 120°

Analysis:

Choice G is correct. The angles inside a quadrilateral total 360°. Angles A and E in the diagram are marked as right angles, therefore they each measure 90°. Angle C is given as 60°, so these three angles together total 240°: 90 ° + 90° + 60° = 240°.

That means Angle ABC must be 120°: 360° − 240° = 120°.

Since the question states that Angle ABC is divided into 2 congruent parts, Angle ABD must be half of Angle ABC: 120° ÷ 2 = 60°.

Choices F and H appear to be guesses and are incorrect. Choice I is incorrect, because Angle ABC measures 120°, but the question asks for the measure of Angle ABD which is half of Angle ABC.

Question **19** *assesses:*

Strand C: Geometry and Spatial Sense

> **Standard 2: The student visualizes and illustrates ways in which shapes can be combined, subdivided, and changed.**
>
> > **MA.C.2.3.1** Understands the geometric concepts of symmetry, reflections, congruency, similarity, perpendicularity, parallelism, and transformations, including flips, slides, turns, and enlargements. (Also assesses B.1.3.3, C.1.2.1, C1.3.1, and C.3.3.1) **(MC)**

Teaching Tip

Draw and label a coordinate grid on the blackboard. Place a variety of shapes with labeled vertices along the board's ledge. Each shape should have a magnet or a piece of tape on one side so it will stick to the board. Ask two students to stand at the board, tell them which shape to choose, and where to place it by naming the coordinates of a vertex. Next, give instructions for a transformation (e.g., reflect the parallelogram over the x-axis). You could also ask students to place a congruent figure in another quadrant. Each instruction should reinforce the concepts tested by the benchmark.

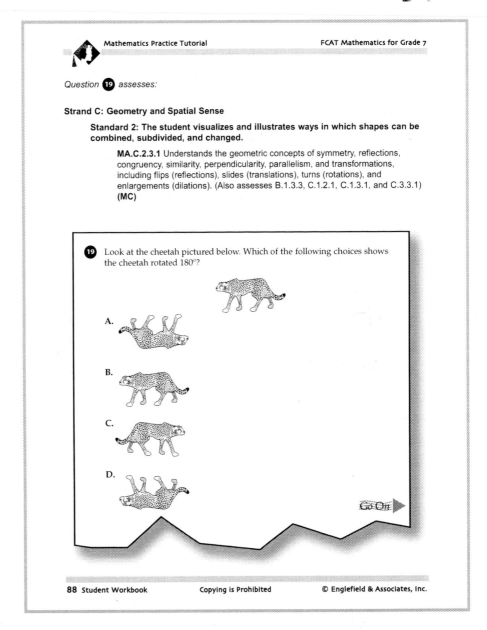

Mathematics Practice Tutorial   FCAT Mathematics for Grade 7

Question **19** *assesses:*

Strand C: Geometry and Spatial Sense

Standard 2: The student visualizes and illustrates ways in which shapes can be combined, subdivided, and changed.

MA.C.2.3.1 Understands the geometric concepts of symmetry, reflections, congruency, similarity, perpendicularity, parallelism, and transformations, including flips (reflections), slides (translations), turns (rotations), and enlargements (dilations). (Also assesses B.1.3.3, C.1.2.1, C.1.3.1, and C.3.3.1) **(MC)**

19 Look at the cheetah pictured below. Which of the following choices shows the cheetah rotated 180°?

A.

B.

C.

D.

Go On ▶

88 Student Workbook   Copying is Prohibited   © Englefield & Associates, Inc.

Analysis:

Choice A is correct. A 180° rotation means that the picture is turned halfway around. Choice B is incorrect because it represents a rotation of 360°, or possibly a translation (slide) to a new location without a change in orientation. Choice C is incorrect because it represents a reflection over a vertical line. Choice D is incorrect because it represents a reflection over a horizontal line.

Question **20** *assesses:*

Strand C: Geometry and Spatial Sense

Standard 2: The student visualizes and illustrates ways in which shapes can be combined, subdivided, and changed.

MA.C.2.3.2 Predicts and verifies patterns involving tessellations (a covering of a plane with congruent copies of the same pattern with no holes and no overlaps, like floor tiles). **(MC)**

Teaching Tip

A tessellation is a tiling or a pattern formed by placing congruent figures together on a surface with no empty spaces or overlapping areas. An example of a tessellation is a checkerboard. Tessellations are fun, but some experience with them is required to fully understand which shapes or combinations of shapes will tessellate. Using manipulatives such as pattern blocks or cardboard cut-outs of polygons is the best way to understand tessellations. However, if your class doesn't have access to manipulatives or don't have time to make them, there are several very good sites on the Web that allow your students to construct tessellations with "virtual manipulatives." Two of these are listed below, but you can find more by typing "interactive tessellations" into your favorite search engine.

The Shodor Education Foundation, Inc. site at
http://www.shodor.org/interactivate/activities/tessellate/

National Library of Virtual Manipulatives hosted by Utah State University at
http://nlvm.usu.edu/en/nav/category_g_2_t_3.htm

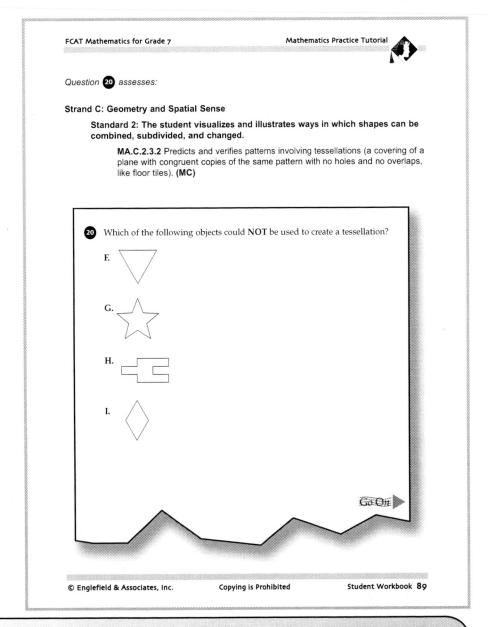

FCAT Mathematics for Grade 7   Mathematics Practice Tutorial

Question **20** *assesses:*

Strand C: Geometry and Spatial Sense

> **Standard 2: The student visualizes and illustrates ways in which shapes can be combined, subdivided, and changed.**
>
> > **MA.C.2.3.2** Predicts and verifies patterns involving tessellations (a covering of a plane with congruent copies of the same pattern with no holes and no overlaps, like floor tiles). **(MC)**

20 Which of the following objects could **NOT** be used to create a tessellation?

F.

G.

H.

I.

Go On ▶

Analysis:

Choice G is correct. A tessellation is a pattern formed by placing congruent figures together with no empty space or overlapping areas. Think of it as a tiling. The tiles on a floor are one kind of tessellation. The star in Choice G is the only shape shown here which cannot form a tiling since it creates gaps and overlaps. The other choices are incorrect because all of them can be used to form tessellations. See the diagrams below.

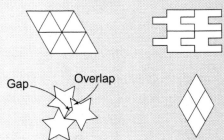

Gap   Overlap

Question **21** *assesses:*

Strand C: Geometry and Spatial Sense

Standard 3: The student uses coordinate geometry to locate objects in both two and three dimensions and to describe objects algebraically.

MA.C.3.3.1 Represents and applies geometric properties and relationships to solve real-world and mathematical problems. (Also assesses C.2.3.1, C.2.3.2, and C.3.2.2) **(MC)**

Teaching Tip

This activity will help students learn how to locate objects on a coordinate grid and teach them how to perform geometric transformations. Draw a coordinate grid on a piece of graph paper and plot out a simple shape such as a triangle, square, or hexagon, etc., labeling its points with letters. Include instructions on how to transform the shape. Students follow the instructions to plot out the new coordinates, label the new points with a prime symbol ('), and connect the new points with lines. For example, if the old point is labeled A, then the transformed point is labeled A'. Transformation instructions may include statements such as:

1. Move each point 3 units right and 2 units up to form a new point. Label the new points with a prime symbol ('), and connect the new points with lines.

2. Add 3 to each *x*-coordinate and subtract 5 from each *y*-coordinate to obtain new points. Label the new points with a prime symbol ('), and connect the new points with lines.

3. Make new points by reversing the coordinates of each old point. That is, take the *x*-coordinate of an old point and make it the *y*-coordinate of a new point and take the *y*-coordinate of an old point and make it the *x*-coordinate of a new point. For example, if an old point has the coordinates (-7, 2), the new point has the coordinates (2, -7). Label the new points with a prime symbol ('), and connect the new points with lines.

4. Transform each old point by multiplying both its *x*-coordinate and its *y*-coordinate by 2 to form a new point. Label the new points with a prime symbol ('), and connect the new points with lines.

Many other transformations can be described either algebraically or with directions. Discuss what effect each transformation has on the shape and try to categorize it as a reflection, translation, rotation, dilation, or combination of effects.

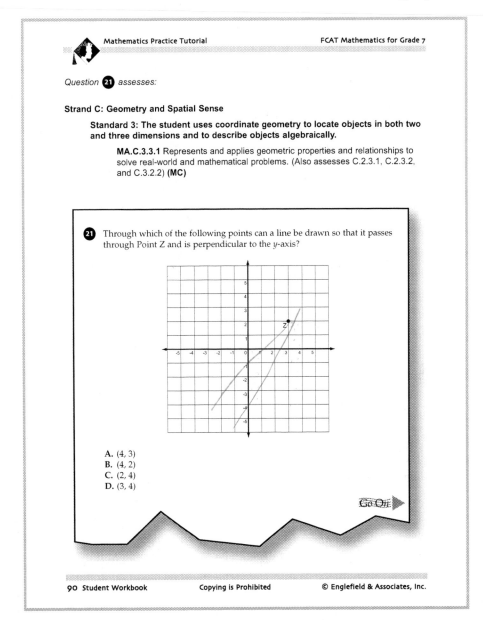

90 Student Workbook   Copying is Prohibited   © Englefield & Associates, Inc.

Analysis:

Choice B is correct. Any line perpendicular to the y-axis must also be parallel to the x-axis and any line parallel to the x-axis must be composed of points that all have the same y-value. For example, a line drawn through the points (0, 1), (1, 1), (2, 1), (3, 1), (4, 1), (5, 1), (x, 1), where x is any real number, is parallel to the x-axis. In order to be parallel to the x-axis and go through point (3, 2), the line must be composed of points that all have 2 as a y-value. Choice B is the only point that has a 2 as its y-value so Choice B is correct. Choices A and C are incorrect because a line passing through their points can be perpendicular to the y-axis or go through Point Z, but not both. Choice D is incorrect because a line passing through (3, 4) can be perpendicular to the x-axis and pass through Point Z, but it cannot be perpendicular to the y-axis and pass through Point Z.

Question **22** *assesses:*

Strand C: Geometry and Spatial Sense

Standard 3: The student uses coordinate geometry to locate objects in both two and three dimensions and to describe objects algebraically.

MA.C.3.3.2 Identifies and plots ordered pairs in all four quadrants of a rectangular coordinate system (graph) and applies simple properties of lines. **(MC)**

Teaching Tip

Create a coordinate grid with four quadrants. Throughout the four quadrants, place 26 points labeled A-Z. Then have students use the coordinates of each letter to spell certain words, sentences, riddles, or messages in this code.

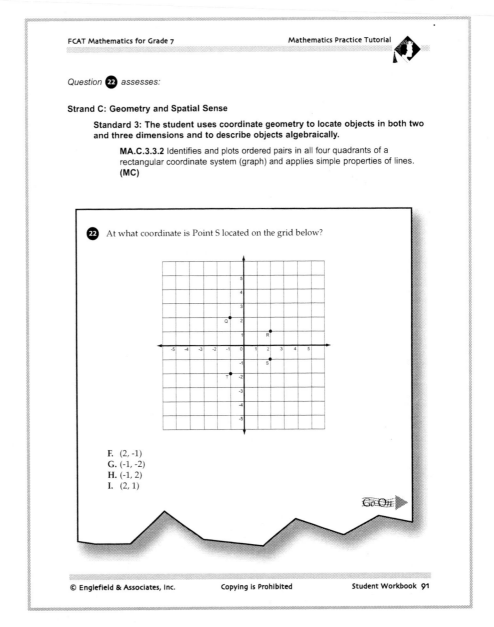

Question **22** *assesses:*

Strand C: Geometry and Spatial Sense

> **Standard 3: The student uses coordinate geometry to locate objects in both two and three dimensions and to describe objects algebraically.**
>
> > **MA.C.3.3.2** Identifies and plots ordered pairs in all four quadrants of a rectangular coordinate system (graph) and applies simple properties of lines. **(MC)**

22 At what coordinate is Point S located on the grid below?

F. (2, -1)
G. (-1, -2)
H. (-1, 2)
I. (2, 1)

Go On ▶

Analysis:

Choice F is correct. Think of points on a coordinate grid as having two names, a first name and a last name. A point's first name is its *x*-coordinate which tells how far right or left of the origin (0, 0) it is. Its last name is its *y*-coordinate which tells how far above or below the origin it is. Point S is 2 units right of the origin and 1 unit below it, so its coordinates are (2, -1). Choice G is incorrect because (-1, -2) are the coordinates of Point T, not Point S. Choice H is incorrect because (-1, 2) are the coordinates of Point Q, not Point S. Choice I is incorrect because (2, 1) are the coordinates of Point R, not Point S.

Question **23** *assesses:*

Strand D: Algebraic Thinking

> **Standard 1: The student describes, analyzes, and generalizes a wide variety of patterns, relations, and functions.**
>
> > **MA.D.1.3.1** Describes a wide variety of patterns, relationships, and functions through models, such as manipulatives, tables, graphs, expressions, equations, and inequalities. (Also assesses A.5.3.1) **(MC, GR)**

Teaching Tip

Try to make students aware of the wide variety of patterns in mathematics and the world at large. Especially take time to investigate nonnumerical patterns. Projects involving tessellations or radial symmetry are a good place to start. These are types of visual patterns. Wooden or plastic polygonal pattern blocks can be arranged in pleasing starbursts of alternating colors. This should appeal to tactile and visually oriented learners. There are also several videos about fractal geometry which students may appreciate for their beauty and visual appeal. Fractals exhibit a unique type of pattern called self-similarity. Students are also usually fascinated by palindromic numbers and words. Palindromes are numbers, words, phrases, and/or sentences that read the same backwards or forwards.

Examples: I prefer pi;
Evil olive

Have them search the Internet to find other amusing examples. Most patterns are not numerical. They can be perceived through any of our senses and most students are not aware of the importance of patterns in human thought. The human brain strives to make sense of the world by pattern recognition and attempting to reconcile new patterns within a broad conceptual framework.

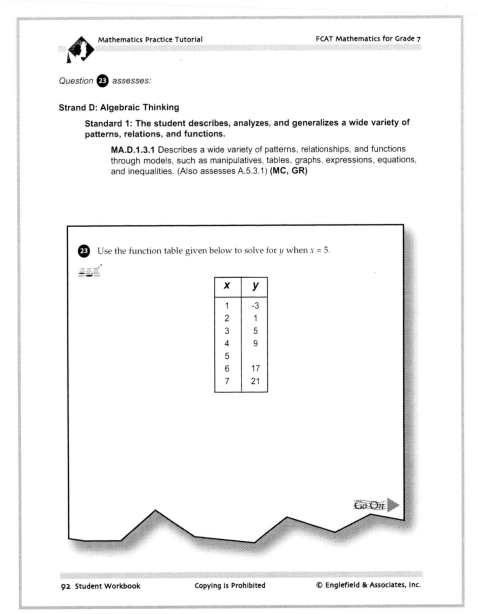

Mathematics Practice Tutorial   FCAT Mathematics for Grade 7

Question **23** *assesses:*

Strand D: Algebraic Thinking

> **Standard 1: The student describes, analyzes, and generalizes a wide variety of patterns, relations, and functions.**
>
> > **MA.D.1.3.1** Describes a wide variety of patterns, relationships, and functions through models, such as manipulatives, tables, graphs, expressions, equations, and inequalities. (Also assesses A.5.3.1) **(MC, GR)**

23 Use the function table given below to solve for y when $x = 5$.

x	y
1	-3
2	1
3	5
4	9
5	
6	17
7	21

Go On ▶

Analysis:

The correct answer is 13. Each time the x-value increases by 1, the y-value increases by 4:

$$-3 + 4 = -1; \ 1 + 4 = 5; \ 5 + 4 = 9; \ 9 + 4 = 13; \ 13 + 4 = 17; \ 17 + 4 = 21.$$

Since the value of y when $x = 4$ is 9, the value of y when $x = 5$ must be 4 more than this:

$$9 + 4 = 13.$$

Question **24** *assesses:*

Strand D: Algebraic Thinking

> **Standard 1: The student describes, analyzes, and generalizes a wide variety of patterns, relations, and functions.**
>
> > **MA.D.1.3.2** Creates and interprets tables, graphs, equations, and verbal descriptions to explain cause-and-effect relationships. (Also assesses A.5.3.1) **(MC, GR)**

Teaching Tip

Use the classroom as a human grid where each student is represented by an ordered pair. Label each row and column like a coordinate. Then, ask students to identify at which point on the "graph" they are located. For example, taking a student's desk position as the *x*-coordinate and the row number as the *y*-coordinate, the student sitting in the third seat of the first row would be identified as (3, 1), while the student in the first seat of the third row is (1, 3). To make it more challenging, choose one student's seat to be the origin of the human graph, then ask students to identify at which points they are located in reference to the origin. In this instance, some students will have negative numbers in their ordered pair. Help students identify their ordered pair by reminding them they can determine whether or not their coordinates are positive or negative by the quadrant in which they are sitting.

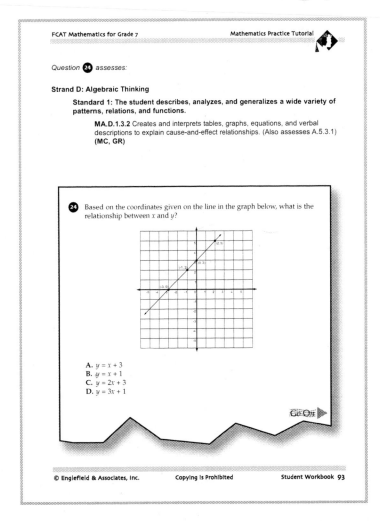

Question **24** *assesses:*

Strand D: Algebraic Thinking

    Standard 1: The student describes, analyzes, and generalizes a wide variety of patterns, relations, and functions.

        MA.D.1.3.2 Creates and interprets tables, graphs, equations, and verbal descriptions to explain cause-and-effect relationships. (Also assesses A.5.3.1) **(MC, GR)**

24 Based on the coordinates given on the line in the graph below, what is the relationship between x and y?

    A. $y = x + 3$
    B. $y = x + 1$
    C. $y = 2x + 3$
    D. $y = 3x + 1$

Analysis:

Choice A is correct. Each y-value is 3 greater than its corresponding x-value. This is represented in the equation $y = x + 3$. If you don't notice this relationship, it's still possible to answer this question by substituting the points on the graph into each equation to see which one they "satisfy." A point "satisfies" an equation if, when its coordinates are substituted into the equation, it creates a true mathematical sentence. If even one of the points on the line does not satisfy the equation, then it is not the right one. Substituting in the four identified points on the graph into the equation $y = x + 3$:

    $(-3, 0)$ $0 = -3 + 3$ true; $(-1, 2)$ $2 = -1 + 3$ true; $(0, 3)$ $3 = 0 + 3$ true; $(2, 5)$ $5 = 2 + 3$ true.

Since all of these points satisfy this equation, this is the correct representation of the graph. Choice B is incorrect because $(-3, 0)$ does not satisfy the equation $y = x + 1$:

    $(-3, 0)$; $0 = -3 + 1$; $0 = -2$; false.

Choice C is incorrect because $(-3, 0)$ does not satisfy the equation $y = 2x + 3$:

    $(-3, 0)$; $0 = 2 \times -3 + 3$; $0 = -6 + 3$; $0 = -3$; false.

Choice D is incorrect because $(-3, 0)$ does not satisfy the equation $y = 3x + 1$:

    $(-3, 0)$; $0 = 3 \times -3 + 1$; $0 = -9 + 1$; $0 = -8$; false.

Question **25** *assesses:*

Strand D: Algebraic Thinking

Standard 2: The student uses expressions, equations, inequalities, graphs, and formulas to represent and interpret situations.

MA.D.2.3.1 Represents and solves real-world problems graphically, with algebraic expressions, equations, and inequalities. (Also assesses A.1.3.3) **(MC)**

Teaching Tip

Make a large copy of the key word chart below on posterboard. Display it in a prominent location and refer to it often when reviewing word problems with the class.

Key Words in Word Problems

Addition	Subtraction	Multiplication	Division	Equals
added to	subtracted from	multiplied by	divided by	is
sum	minus	product of	quotient of	are
total of	difference of/between	times	per	was
more than	fewer than	increased or decreased	ratio of	were
increased by	decreased by	by a factor of	out of	will be
combined together	fewer than	of	percent	gives
	less or less than	double – multiply by 2	a (sometimes)	yields
		triple – multiply by 3		sold for

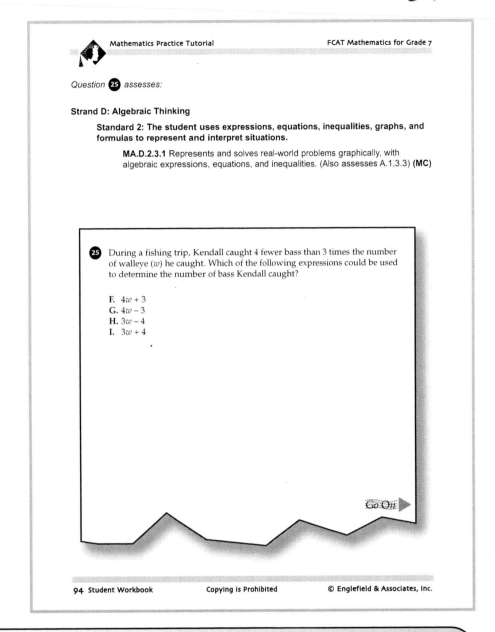

Mathematics Practice Tutorial   FCAT Mathematics for Grade 7

Question **25** *assesses:*

Strand D: Algebraic Thinking

> **Standard 2: The student uses expressions, equations, inequalities, graphs, and formulas to represent and interpret situations.**
>
> > **MA.D.2.3.1** Represents and solves real-world problems graphically, with algebraic expressions, equations, and inequalities. (Also assesses A.1.3.3) **(MC)**

25 During a fishing trip, Kendall caught 4 fewer bass than 3 times the number of walleye (w) he caught. Which of the following expressions could be used to determine the number of bass Kendall caught?

F. $4w + 3$
G. $4w - 3$
H. $3w - 4$
I. $3w + 4$

94 Student Workbook   Copying is Prohibited   © Englefield & Associates, Inc.

Analysis:

Choice H is correct. Questions like this test your students' ability to take a written statement and convert it into a mathematical expression. It is important that they understand what the "key" words mean. The question says Kendall caught 4 fewer bass than 3 times the number of walleye he caught. The phrase "3 times the number of walleye" is easily translated as 3 x w or $3w$. The word "fewer" means something is less than something else which is interpreted mathematically as subtraction, in this case as -4. Combining these two expressions means that the number of bass Kendall caught is expressed as $3w - 4$. You can check your work by creating an imaginary situation trying your expression. Suppose Kendall caught 3 walleye. According to the expression, that means he caught $3(3) - 4 = 5$ bass. Does this match what the question tells you? Yes, 5 bass is 4 less than 3 times the number of walleye. Choice F is incorrect because the expression $4w + 3$ matches the statement "Kendall caught 3 more bass than 4 times the number of walleye he caught," not the phrase stated in the question. Choice G is incorrect because the expression $4w - 3$ matches the statement "Kendall caught 3 fewer bass than 4 times the number of walleye he caught," not the phrase stated in the question. Choice I is incorrect because the expression $3w + 4$ matches the statement "Kendall caught 4 more bass than 3 times the number of walleye he caught," not the phrase stated in the question.

Question **26** *assesses:*

Strand D: Algebraic Thinking

Standard 2: The student uses expressions, equations, inequalities, graphs, and formulas to represent and interpret situations.

MA.D.2.3.2 Uses algebraic problem-solving strategies to solve real-world problems involving linear equations and inequalities. **(MC, GR)**

Teaching Tip

Introduce students to the concept of solving a formula for one of the other unknowns. For example, start with a familiar formula like $A = lw$ and show students that it's equivalent to both $l = A \div w$ and $w = A \div l$. Give them several different problems that require them to find each of the three unknowns, A, l, and w. Emphasize the fact that if a formula contains 3 parameters like $A = lw$ or $d = rt$, 2 must be known. If it contains 5 parameters, 4 must be known, etc. After reviewing several formulas that require one-step conversions, try harder multi-step conversions such as solving for °F in the formula.

$$°C = \frac{5}{9} \times (°F - 32)$$

Note: It doesn't really matter if students can do the multi-step conversions. The important thing at this grade level is that they recognize that any formula can be used to solve for any of its parameters by converting it with the proper order of operations.

© Englefield & Associates, Inc.

Question **26** *assesses:*

Strand D: Algebraic Thinking

> **Standard 2: The student uses expressions, equations, inequalities, graphs, and formulas to represent and interpret situations.**
>
> > **MA.D.2.3.2** Uses algebraic problem-solving strategies to solve real-world problems involving linear equations and inequalities. **(MC, GR)**

26 Mario spends a total of $163.00 on doughnuts for a school fundraiser breakfast at a local bakery that sells doughnuts by the dozen only. The first dozen cost $10.00 and each additional dozen costs $8.50. This information is displayed in the equation below, where n represents the number of dozens of doughnuts Mario bought.

$$\$10.00 + (n - 1)(\$8.50) = \$163.00$$

According to this equation, how many dozens of doughnuts did Mario buy?

Go On ▶

Analysis:

The correct answer is 19 dozen doughnuts. Begin by distributing (multiplying) the $8.50 by the terms in the parentheses:

$$\$10.00 + (n - 1)(\$8.50) = \$163.00; \ \$10.00 + \$8.50n - \$8.50 = \$163.00.$$

Next collect like terms by subtracting:

$$\$10.00 - \$8.50: \$1.50 + \$8.50n = \$163.00.$$

Next, subtract $1.50 from each side of the equation:

$$\$1.50 + \$8.50n - \$1.50 = \$163.00 - \$1.50; \ \$8.50n = \$161.50.$$

Finally, divide each side of the equation by $8.50:

$$\$8.50n \div \$8.50 = \$161.50 \div \$8.50$$
$$n = 19 \text{ dozen.}$$

Question **27** *assesses:*

Strand E: Data Analysis and Probability

Standard 1: The student understands and uses the tools of data analysis for managing information.

MA.E.1.3.1 Collects, organizes, and displays data in a variety of forms, including tables, line graphs, charts, and bar graphs, to determine how different ways of presenting data can lead to different interpretations. (Also assesses E.1.3.3) **(MC, GR)**

Teaching Tip

Measure the height of every student in the classroom in inches and place the data in a table. Then, create different types of graphs to display the data. For example, first use a stem-and-leaf plot. Remind students that in a stem-and-leaf plot, the greatest common place value of the data is used to form the "stem." The numbers on the next greatest place value position are then used to form the "leaves." Then, use the same data to construct a bar graph and a pie chart. What special things do you have to consider when creating bar graphs and pie charts? (For example, it is probably both necessary and desirable to group students together into "height intervals" so that each student doesn't have his or her own bar or section of the pie.) How do the graphs differ? What are the strengths and weaknesses of each type of graph? What information can be obtained from each type of graph?

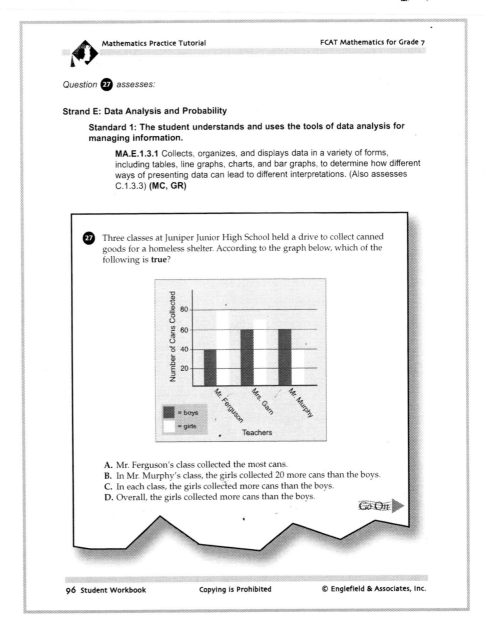

Question **27** assesses:

Strand E: Data Analysis and Probability

> **Standard 1: The student understands and uses the tools of data analysis for managing information.**
>
> > **MA.E.1.3.1** Collects, organizes, and displays data in a variety of forms, including tables, line graphs, charts, and bar graphs, to determine how different ways of presenting data can lead to different interpretations. (Also assesses C.1.3.3) **(MC, GR)**

27 Three classes at Juniper Junior High School held a drive to collect canned goods for a homeless shelter. According to the graph below, which of the following is **true**?

A. Mr. Ferguson's class collected the most cans.
B. In Mr. Murphy's class, the girls collected 20 more cans than the boys.
C. In each class, the girls collected more cans than the boys.
D. Overall, the girls collected more cans than the boys.

Go On▶

Analysis:

Choice D is correct. According to the chart, the girls in all 3 classes collected a total of 190 cans (80 + 70 + 40 = 190) while the boys only collected 160 cans (40 + 60 + 60 = 160). Choice A is incorrect because Mrs. Garn's class collected 130 cans (60 + 70 = 130) while Mr. Ferguson's class only collected 120 cans (40 + 80 = 120). Choices B and C are incorrect, because in Mr. Murphy's class the boys collected more cans than the girls.

Question **28** *assesses:*

Strand E: Data Analysis and Probability

Standard 1: The student understands and uses the tools of data analysis for managing information.

MA.E.1.3.2 Understands and applies the concepts of range and central tendency (mean, median, and mode). (Also assesses E.1.3.3) **(MC, GR)**

Teaching Tip

Students may work individually, in pairs, or in small groups for this activity. Shuffle together one or more standard decks of cards. Distribute a series of cards to each student or group. Reading each ace as a 1, each face card as a 10, and all other cards as the number they display, ask students to find the range, median, mode, and mean of their data set. Give each student or group three additional cards, and ask them how each of the measures of central tendency and the range are affected. Decks with different values may also be created using 3" x 5" index cards.

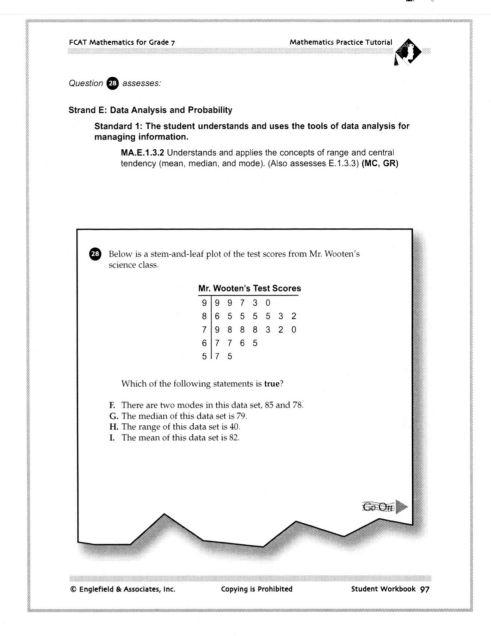

Question **28** *assesses:*

Strand E: Data Analysis and Probability

> **Standard 1: The student understands and uses the tools of data analysis for managing information.**
>
>> **MA.E.1.3.2** Understands and applies the concepts of range and central tendency (mean, median, and mode). (Also assesses E.1.3.3) **(MC, GR)**

28 Below is a stem-and-leaf plot of the test scores from Mr. Wooten's science class.

Mr. Wooten's Test Scores

9	9	9	7	3	0		
8	6	5	5	5	5	3	2
7	9	8	8	8	3	2	0
6	7	7	6	5			
5	7	5					

Which of the following statements is **true**?

F. There are two modes in this data set, 85 and 78.
G. The median of this data set is 79.
H. The range of this data set is 40.
I. The mean of this data set is 82.

Go On ▶

Analysis:

Choice G is correct. Since 79 is the 13th term in a set of 25, it is the middle term and therefore the median. Choice F is incorrect because there is only one mode, 85. If there had been four 78s instead of three 78s would there have been two modes. Choice H is incorrect because the range of this data set is 44 (99 – 55 = 44), not 40. Choice I is incorrect because the mean of this data set is 78.96, not 82.

Question **29** *assesses:*

Strand E: Data Analysis and Probability

Standard 1: The student understands and uses the tools of data analysis for managing information.

MA.E.1.3.3 Analyzes real-world data by applying appropriate formulas for measures of central tendency and organizing data in a quality display, using appropriate technology, including calculators and computers. **(MC, GR)**

Teaching Tip

Organize the students in your classroom to collect data about each other on a number of different measures. These may include: height, weight, shoe size, eye color, birth month, birth state, etc. Students may suggest more. Measure as many of these attributes as possible right in the classroom and discuss the best way to measure each attribute. For example, students will probably know how to measure height and weight, but they may need some help measuring shoe size. Several sites on the Web explain how to convert a foot's length in inches or centimeters to shoe size. Allow students a chance to suggest solutions to these problems. Discuss which of the measures of central tendency and range is applicable to each data set. For example, the concepts of mean, median, mode, and range all have some meaning when applied to the height and weight data sets, but mode may be the only one applicable to eye color or birth state. Discuss how some of the non-numerical data can be converted to numerical data. For example, for birth month, January could be represented by a 1, February by a 2, March by a 3, etc. Discuss what type of graph is most appropriate for displaying each data set, then ask students to construct the graphs.

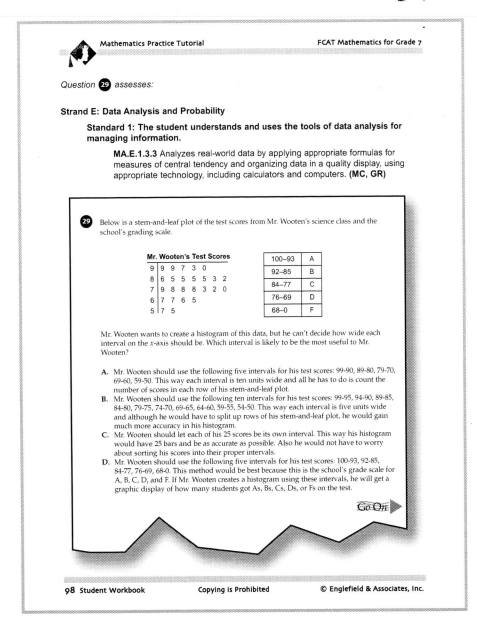

Question **29** *assesses:*

Strand E: Data Analysis and Probability

Standard 1: The student understands and uses the tools of data analysis for managing information.

MA.E.1.3.3 Analyzes real-world data by applying appropriate formulas for measures of central tendency and organizing data in a quality display, using appropriate technology, including calculators and computers. **(MC, GR)**

29 Below is a stem-and-leaf plot of the test scores from Mr. Wooten's science class and the school's grading scale.

Mr. Wooten's Test Scores

9	9 9 7 3 0
8	6 5 5 5 5 3 2
7	9 8 8 8 3 2 0
6	7 7 6 5
5	7 5

100–93	A
92–85	B
84–77	C
76–69	D
68–0	F

Mr. Wooten wants to create a histogram of this data, but he can't decide how wide each interval on the *x*-axis should be. Which interval is likely to be the most useful to Mr. Wooten?

A. Mr. Wooten should use the following five intervals for his test scores: 99-90, 89-80, 79-70, 69-60, 59-50. This way each interval is ten units wide and all he has to do is count the number of scores in each row of his stem-and-leaf plot.

B. Mr. Wooten should use the following ten intervals for his test scores: 99-95, 94-90, 89-85, 84-80, 79-75, 74-70, 69-65, 64-60, 59-55, 54-50. This way each interval is five units wide and although he would have to split up rows of his stem-and-leaf plot, he would gain much more accuracy in his histogram.

C. Mr. Wooten should let each of his 25 scores be its own interval. This way his histogram would have 25 bars and be as accurate as possible. Also he would not have to worry about sorting his scores into their proper intervals.

D. Mr. Wooten should use the following five intervals for his test scores: 100-93, 92-85, 84-77, 76-69, 68-0. This method would be best because this is the school's grade scale for A, B, C, D, and F. If Mr. Wooten creates a histogram using these intervals, he will get a graphic display of how many students got As, Bs, Cs, Ds, or Fs on the test.

Go On ▶

Analysis:

Choice D is correct. Decisions about the scales on a graph or its appearance should be made by considering its purpose. A teacher is probably most interested in the number of students getting As, Bs, and so on, so using the school's grading scale for the width of his intervals makes the most sense. Choices A, B, and C are incorrect because they do not consider the purpose of the graph. Also, intervals do not always have to be the same width. It depends on what they are trying to show. Choice C also has another problem. Each test score would not have its own interval because there are many duplicate scores in this data set. For instance, there are two 99s, four 85s, three 78s, and two 67s.

Question **30** *assesses:*

Strand E: Data Analysis and Probability

Standard 2: The student identifies patterns and makes predictions from an orderly display of data using concepts of probability and statistics.

MA.E.2.3.1 Compares experimental results with mathematical expectations of probabilities. **(MC)**

Teaching Tip

First explain the concepts of theoretical (calculated) probability versus experimental probability. If you attempt any probability experiment, you will soon find a difference between the outcomes you find experimentally and the outcomes predicted by probability calculations. Why is this true? Because, given enough time and opportunity, anything that can happen, will happen—eventually. Activity: Using a standard deck of cards, explain how to calculate the probability of drawing a diamond (13/52 = 1/4). Explain that the probability of drawing any one of the other suits is exactly the same. Then do a probability experiment with the whole class. Each child should choose one card at random from the deck, announce the suit drawn, replace the card, shuffle the deck, and pass it on to the next student. A record of each card's suit should be kept on the board. After each student has had a turn, discuss the difference between experimental probability and theoretical probability. A good extension of this discussion can be made on predicting the weather. For example, a 20% chance of rain may not seem like a very good chance, but what it means is that in 20 times out of a hundred when weather conditions are the same, it rains. Often students will say that the local weather person got the forecast wrong if it rains on a 20% chance. This is a misunderstanding of probability and weather forecasting methodology.

Question **30** *assesses:*

Strand E: Data Analysis and Probability

> **Standard 2: The student identifies patterns and makes predictions from an orderly display of data using concepts of probability and statistics.**
>
> > **MA.E.2.3.1** Compares experimental results with mathematical expectations of probabilities. **(MC)**

30 Abe flipped a penny 5 times. Each time, it landed on "heads." Which of the following statements is **true**?

F. The next time Abe flips the penny, it will most likely land on "heads."
G. The next time Abe flips the penny, it will most likely land on "tails."
H. The penny will most likely land on "tails" the next 5 times it is flipped.
I. The next time Abe flips the penny, it has an equal chance of landing on "heads" or "tails."

Go On ▶

Analysis:

Choice I is correct. Every flip of the coin is an independent event. Regardless of what has happened on previous flips, the chances of an honest coin landing on "heads" is always equal to the chance that it will land on "tails." Because of this, all of the other choices are incorrect.

Question **31** *assesses:*

Strand E: Data Analysis and Probability

Standard 2: The student identifies patterns and makes predictions from an orderly display of data using concepts of probability and statistics.

MA.E.2.3.2 Determines odds for and odds against a given situation. (Also assesses E.2.2.2) **(MC)**

Teaching Tip

Odds are not the same as probability. *Odds for an event* is defined as the ratio of the number of favorable outcomes to the number of unfavorable outcomes; represented as a fraction or a ratio:

$$\text{Odds for E} = \frac{\text{number of favorable outcomes}}{\text{number of unfavorable outcomes}}$$

Contrast this with the *probability of an event* which is defined as:

$$\text{Probability of E} = \frac{\text{number of favorable outcomes}}{\text{total number of outcomes}}$$

Although odds can be expressed as a fraction, they are usually expressed as a horizontal ratio, *x:y*. Develop a list of simple probability problems and have students find both the probabilities of and the odds for the same event. Allow students to work in pairs or teams and be sure to compare and contrast results in a whole class discussion.

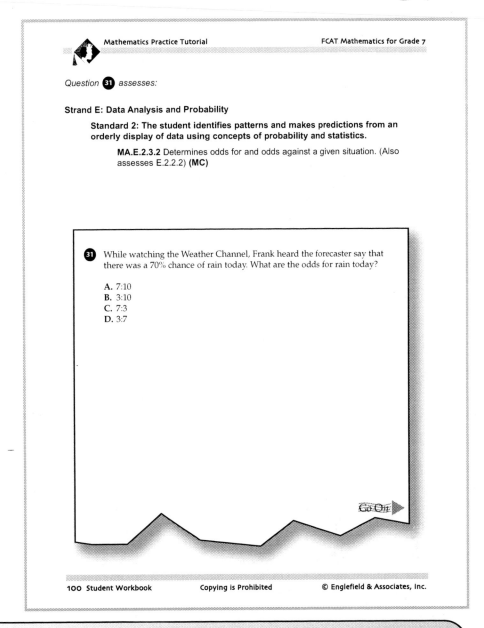

Mathematics Practice Tutorial   FCAT Mathematics for Grade 7

Question **31** *assesses:*

Strand E: Data Analysis and Probability

> **Standard 2: The student identifies patterns and makes predictions from an orderly display of data using concepts of probability and statistics.**
>
> > **MA.E.2.3.2** Determines odds for and odds against a given situation. (Also assesses E.2.2.2) **(MC)**

31 While watching the Weather Channel, Frank heard the forecaster say that there was a 70% chance of rain today. What are the odds for rain today?

A. 7:10
B. 3:10
C. 7:3
D. 3:7

Go On ▶

100 Student Workbook   Copying is Prohibited   © Englefield & Associates, Inc.

Analysis:

Choice C is correct. A 70% chance of rain is the probability that it will rain today. What it means is that in 70 days out of a hundred when weather conditions are the same as they are today, it rains. It also means that 30 days out of a hundred when weather conditions are the same as they are today, it does not rain. Also, remember that odds and probability are not the same thing. The odds for an event is defined as the ratio of the number of favorable outcomes to the number of unfavorable outcomes (in this case, number of days it rains with similar conditions : number of days it doesn't rain with similar conditions). Odds can be expressed as a fraction, but they are usually expressed as a horizontal ratio, *x:y*. Since 70 days out of 100 with the same conditions it rains, and 30 days out of 100 it doesn't rain, the odds for rain today are 70:30 which reduces to 7:3. Choice A is incorrect because it represents the probability that it will rain today, not the odds for rain. Choice B is incorrect because it represents the probability of not raining today, not the odds for rain. Choice D is incorrect because it represents odds against rain today, not the odds for rain.

Question **32** *assesses:*

Strand E: Data Analysis and Probability

Standard 3: The student uses statistical methods to make inferences and valid arguments about real-world situations.

MA.E.3.3.1 Formulates hypotheses, designs experiments, collects and interprets data, and evaluates hypotheses by making inferences and drawing conclusions based on statistics (range, mean, median, and mode) and tables, graphs, and charts. (Also assesses E.3.3.2) **(MC)**

Teaching Tip

Help students gain confidence at collecting and organizing data by having small groups of them perform an actual survey or experiment in which they collect a sizable amount of data that they must then organize and interpret. Provide the following checklist for students to refer to while doing this project.

☐ I used an adequate sample size.

☐ I completed graphs properly.

☐ I generalized the results.

☐ I interpreted the numerical data correctly.

☐ I calculated the range, mean, median, and mode to represent the data collected.

Once the project is completed, have each group present its findings. Ask other students to comment on the presenting group's survey, procedure, and findings.

Question **32** *assesses:*

Strand E: Data Analysis and Probability

> **Standard 3: The student uses statistical methods to make inferences and valid arguments about real-world situations.**
>
> > **MA.E.3.3.1** Formulates hypotheses, designs experiments, collects and interprets data, and evaluates hypotheses by making inferences and drawing conclusions based on statistics (range, mean, median, and mode) and tables, graphs, and charts. (Also assesses E.3.3.2) **(MC)**

32 Look at the graph below.

Favorite Pizza Toppings Survey

Which of the following is the best interpretation of this circle graph?

F. The people who did not respond don't like pizza.
G. The people who did not respond like all three toppings and could not decide which to choose.
H. The people who did not respond did not return their surveys.
I. The people who did not respond like something other than cheese, pepperoni, or mushrooms.

Go On ▶

Analysis:

Choice H is correct. Choices F, G, and I are incorrect because if the people did not respond, you have no data about any of their opinions, so you cannot say what toppings they like or don't like or even whether or not they like pizza at all. The only thing you know for sure is that they did not return their surveys, so Choice H is the only possible answer.

Question **33** *assesses:*

Strand E: Data Analysis and Probability

Standard 3: The student uses statistical methods to make inferences and valid arguments about real-world situations.

MA.E.3.3.2 Identifies the common uses and misuses of probability and statistical analysis in the everyday world. **(MC)**

Teaching Tips

• Using the data gathered from the survey described in the previous Teaching Tip, have students examine each others' data to suggest improvements and to look for errors such as faulty logic, misrepresentation of information, and unsupported conclusions.

• Ask students to find ads in magazines, newspapers, and on the Web that they feel present a biased or untruthful message. Select several for whole class discussion. Try to concentrate on those that seem to misuse concepts of probability and statistics. The Federal Trade Commission's Truth-in-Advertising Rules* may help you guide the discussion. They are summarized below:

1. Advertising must be truthful and non-deceptive. It cannot contain a statement or omit information that is likely to mislead reasonable consumers or is important to the customer's decision to use or buy the product. The Federal Trade Commission (FTC) looks at the ad in context: its words, phases, and pictures. The FTC considers both the expressed (stated) claims and the implied (unstated) claims. It also looks for any failure to include information that leaves consumers with a misimpression about the product.

2. Advertisers must have evidence to back up their claims, especially about claims concerning a product's performance, features, safety, price, or effectiveness. Statements from satisfied customers usually are not sufficient. Offering a money-back guarantee is not a substitute for evidence.

3. Advertisements cannot be unfair. They cannot cause or be likely to cause substantial injury or outweigh any benefit to customers.

The FTC pays closest attention to ads that make claims about health or safety such as:

ABC Sunscreen will reduce the risk of skin cancer.

ABC Water Filters remove harmful chemicals from tap water.

ABC Chainsaw's safety latch reduces the risk of injury.

and ads that make claims that consumers would have trouble evaluating for themselves, such as:

ABC Refrigerators will reduce your energy costs by 25%.

ABC Gasoline decreases engine wear.

ABC Hairspray is safe for the ozone.

*Source: Federal Trade Commission, Truth-in-Advertising Rules; http://www.ftc.gov/bcp/conline/pubs/buspubs/ad-faqs.htm

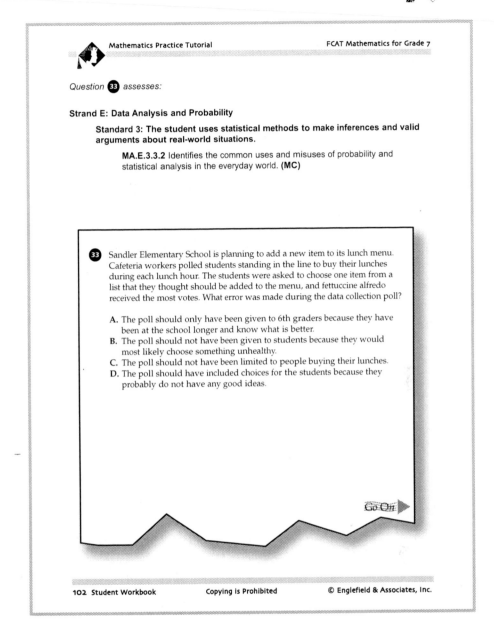

Question **33** *assesses:*

Strand E: Data Analysis and Probability

Standard 3: The student uses statistical methods to make inferences and valid arguments about real-world situations.

MA.E.3.3.2 Identifies the common uses and misuses of probability and statistical analysis in the everyday world. **(MC)**

33 Sandler Elementary School is planning to add a new item to its lunch menu. Cafeteria workers polled students standing in the line to buy their lunches during each lunch hour. The students were asked to choose one item from a list that they thought should be added to the menu, and fettuccine alfredo received the most votes. What error was made during the data collection poll?

A. The poll should only have been given to 6th graders because they have been at the school longer and know what is better.
B. The poll should not have been given to students because they would most likely choose something unhealthy.
C. The poll should not have been limited to people buying their lunches.
D. The poll should have included choices for the students because they probably do not have any good ideas.

Go On ▶

Analysis:

Choice C is correct. The poll should have included all students so there is no bias in the poll and every student's opinion is collected. The students who weren't in the lunch line the day of the poll might buy their lunches other days, and students who don't usually buy their lunches might buy them if the item they want is on the menu. Choice A is incorrect because surveying only 6th graders is likely to introduce bias, and there is no reason to believe that their opinions would be more valuable than anyone else's. Choices B and D are incorrect because they make assumptions about the students' opinions with no supporting data. The whole idea of a survey is to find out what students think. If you already know what they think, the survey is unnecessary.

Mathematics Assessment One Introduction Page as it appears on page 111 in the
Show What You Know® on the 7th Grade FCAT, Mathematics Student Workbook.

Mathematics Assessment One

Directions for Taking the Mathematics Assessment One

On this section of the Florida Comprehensive Assessment Test (FCAT), you will answer 50 questions.

For multiple-choice questions, you will be asked to pick the best answer out of four possible choices and fill in the answer in the answer bubble. On gridded-response questions, you will also fill your answer in answer bubbles, but you will also fill in numbers and symbols corresponding to the solution you obtain for a question. Fill in the answer bubbles and gridded-response answer bubbles on the Answer Sheets on pages 133–137 to mark your selection.

Read each question carefully and answer it to the best of your ability. If you do not know an answer, you may skip the question and come back to it later.

Figures and diagrams with given lengths and/or dimensions are not drawn to scale. Angle measures should be assumed to be accurate. Use the Mathematics Reference Sheet on page 31 to help you answer the questions.

When you finish, check your answers.

MA.B.2.3.2

1   *The Statue of Liberty weighs 408,000 pounds*. How many tons does the Statue of Liberty weigh?*
*Wikipedia; http://en.wikipedia.org/wiki/Statue_of_Liberty

A.   40.8 tons
B.   204 tons
C.   408 tons
D.   816 tons

Analysis: *Choice B is correct.* Since there are 2000 pounds in a ton, divide the weight of the statue in pounds by 2000: 408,000 lbs ÷ 2000 lbs/ton = 204 tons.

MA.A.3.3.3

2   *Pete and Becca own a llama farm that has 240 llamas. Each llama they sell earns them $325.00 profit. If they sell 60% of their llamas, how much total profit will they make?*

F. $31,200.00
G. $46,800.00
H. $62,400.00
I. $78,000.00

Analysis: *Choice G is correct.* First find how many llamas they sell by finding 60% of 240: 60% of 240 = .60 x 240 = 144 llamas. Since they make a profit of $325.00 on each llama they sell, their total profit is $46,800.00: $325.00 x 144 = $46,800.00 profit.

MA.E.3.3.1

3   *The circle graph below displays Wizard Productions' sales figures for the year 2000. Video sales were 10% of total sales and brought in $3,600,000. Use this information to determine how much money book sales earned.*

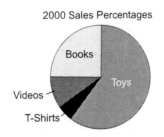

2000 Sales Percentages

A. $7,200,000
B. $9,000,000
C. $14,400,000
D. $16,200,000

Analysis: *Choice B is correct.* We are told that video sales were $3,600,000 and accounted for 10% of total sales. This means that 10% and $3,600,000 are different ways of representing the same amount. We can rephrase this statement to help us find the dollar amount of total sales. We could ask, "10% of what number is $3,600,000?" This question translates into the mathematical statement: 10% of t = $3,600,000 or .10 x t = $3,600,000. To solve .10 x t = $3,600,000 for t, divide both sides of the equation by .10: t = $36,000,000. (Remember, when you divide any number by a fraction or decimal between 0 and 1, the quotient is larger than the number divided.) Now we know that total sales in the year 2000 were $36,000,000. Looking at the circle graph, it's easy to see that book sales appear to be 25% of total sales, so find 25% of $36,000,000: 25% of $36,000,000 = .25 x $36,000,000 = $9,000,000. So, the total dollar amount of sales from books is $9,000,000.

MA.A.3.3.2

4 *Look at the expression given below and determine what should be done first?*

$$(3^2 - 4) \times 2 + 8$$

F. subtract 4 from 3
G. add 2 and 8
H. multiply 4 by 2
I. multiply 3 by 3

Analysis: *Choice I is correct.* Use PEMDAS to help you remember the correct order of operations. The P in PEMDAS stands for parentheses, however, this expression has two operations inside the parentheses. Continuing with the correct order of operations, the E in PEMDAS stands for exponents. The 3^2 is an exponential expression, which means to multiply 3 by itself:

$$3^2 = 3 \times 3$$

MA.C.2.3.1

5 *Which of the following pairs of lines are parallel to one another?*

A.

B.

C.

D.

Analysis: *Choice C is correct.* Parallel lines are two lines in the same plane that never intersect, no matter how far they are extended. It's easy to see that the lines in all of the choices except Choice C would intersect if one or both of them were made longer.

MA.C.1.3.1

6 *What is the sum of the interior angles of this shape? Your answer will be in degrees.*

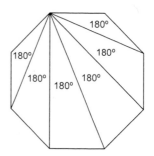

Analysis: *The correct answer is 1080°.* Consider the diagram below.

An octagon can be divided into 6 triangles as shown above. Every triangle has an internal angle sum of 180°, and since the internal angle sum of the octagon is the same as the total angle sums of all of these 6 triangles, an octagon has a total of 1080°:

$$6 \times 180° = 1080°$$

MA.D.1.3.2

7 *Which of the following is NOT true about the line $y = 3$?*
F. It is parallel to the y-axis.
G. It passes through (2, 3).
H. It does not pass through (3, 4).
I. It never crosses the x-axis.

Analysis: *Choice F is correct.* A line that is parallel to the y-axis must be a vertical line, and $y = 3$ is a horizontal line. A vertical line has an equation where x is constant and a y term is not even a part of the equation. For example, $x = 3$ is the equation of a vertical line (parallel to the y-axis) passing through (3, 0), (3, 1), (3, 2), (3, 3), etc. In the equation $x = 3$, it doesn't matter what y is, because y is not a part of the equation, but x is always 3. The other choices are incorrect because they are all true. For the graph of the equation $y = 3$ to pass through a point, the point's y-coordinate must be a 3, so this line does pass through (2, 3) in Choice G and does not pass through (3, 4) in Choice H. Since no horizontal line ever crosses the x-axis (except the x-axis itself), Choice I is also true and hence an incorrect answer.

MA.D.2.3.2

8 *Robin is competing in an archery tournament. On her previous three arrows, she scored 7, 0, and 3 points, respectively. Her total for this round of the tournament is 17 points so far. She has one more arrow to shoot. Which of the following expressions best represents her final point total (p) for the round?*

A. $p > 17$
B. $p < 17$
C. $p = 17$
D. $p \geq 17$

Analysis: *Choice D is correct.* Since Robin already has 17 points, if she shoots a 0 on her next arrow, then her point total, p, will still be 17. If she shoots anything other than a 0, her point total will be greater than 17. Therefore, the best answer is $p \geq 17$.

MA.E.1.3.3

9 *Jay and his friends ordered pizza for their party. The graph below shows how many pieces of pizza each person ate. Use this graph to determine the mode of the number of pieces eaten.*

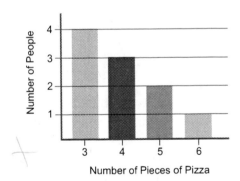

Number of Pieces of Pizza

F. 3 pieces
G. 4 pieces
H. 5 pieces
I. 6 pieces

Analysis: *Choice F is correct.* The mode is the number that occurs most frequently in a set of data. The data in the graph can be displayed as a set of numbers with each number showing how many pieces of pizza one person ate: {3, 3, 3, 3, 4, 4, 4, 5, 5, 6}. The graph shows that 4 people ate 3 pieces of pizza each, 3 people ate 4 pieces, 2 people ate 5 pieces, and one person ate 6 pieces. We are asked to find the mode of the number of pieces eaten, so the mode is 3, since it occurs in the data set most often—4 times.

MA.E.2.3.2

10 *Horatio is reading a book about sunken ships in the Mediterranean Sea. The book is 256 pages long and 64 of the pages have illustrations on them. What are the odds that Horatio will open the book at random to a page with a picture on it?*

A. 1:256
B. 1:64
C. 1:4
D. 1:3

Analysis: *Choice D is correct.* Odds are not the same as probability. The *odds* for an event is defined as the ratio of the number of favorable outcomes to the number of unfavorable outcomes (in this case, number of pages with illustrations : number of pages without illustrations). The number of pages with illustrations is given as 64, so the number of pages without illustrations is 192:

$$256 - 64 = 192$$

The odds for randomly opening this book to a page with a picture on it is 64:192 which reduces to 1:3.

MA.A.3.3.3

11 *Yolanda is helping the drama club construct a set for an upcoming play. They need to construct a giant pencil for a stage prop. They base their design on a real pencil that is 6.5 inches long. If each inch of the real pencil represents 3 feet of the giant pencil, how long will the giant pencil be?*

F. 9.5 feet
G. 13.0 feet
H. 19.5 feet
I. 26.0 feet

Analysis: *Choice H is correct.* Since each inch of the model equals 3 feet on the giant pencil, multiply the length of the model by 3:

$$6.5 \text{ inches} \times 3 \text{ feet/inch} = 19.5 \text{ feet}$$

MA.E.2.3.1

12 *Randy randomly selects 16 pieces of candy from a bag that contains an equal number of red, green, yellow, and blue pieces. Of those 16 pieces of candy, 6 of them are red. How does this result compare with what would be expected based on probability?*

A. The number of pieces of red candy Randy selected is greater than what would be expected.

B. The number of pieces of red candy Randy selected is less than what would be expected.

C. The number of pieces of red candy Randy selected is equal to what would be expected.

D. The expected probability cannot be determined without knowing the exact number of pieces of candy in the bag.

Analysis: *Choice A is correct.* Since each of the four colors in the bag is equally represented, you would expect each color to have an equal chance of being chosen randomly—a 1 in 4 or a 25% chance. In a selection of 16 pieces, you would expect 4 to be red:

$$25\% \text{ of } 16 = 0.25 \times 16 = 4$$

Since Randy got 6 red pieces in his sample, he got more than what is expected based on probability.

MA.B.1.3.3

13 *Kyle is helping fix playground equipment at a local park. He is responsible for increasing the size of the sandbox, shown below. To do this, he will add 2 feet to the width of the sandbox. How will the perimeter of the sandbox change if Kyle adds 2 feet to the width of the sandbox?*

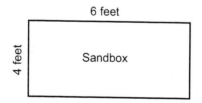

F. The perimeter will increase by 2 feet.

G. The perimeter will increase by 4 feet.

H. The perimeter will increase by 8 feet.

I. The perimeter will be twice as large.

Analysis: *Choice G is correct.* The perimeter of the current sandbox is 20 feet:

$$4 + 6 + 4 + 6 = 20$$

If 2 feet is added to the width of the sandbox, the new dimensions of the sandbox will be 6 feet wide by 6 feet long. This means that the new perimeter will be 24 feet:

$$6 + 6 + 6 + 6 = 24$$

so the perimeter will increase by 4 feet.

MA.C.3.3.2

14 *Consuela used a grid to map out some of her favorite places in town. What is located at the point (-2, 4)?*

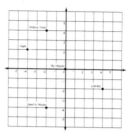

A. the park

B. the library

C. the willow tree

D. Dawn's house

Analysis: *Choice C is correct.* Think of points on a coordinate grid as having two names, a first name and a last name. A point's first name is its x-coordinate which tells how far right or left of the origin (0, 0) it is. Its last name is its y-coordinate which tells how far above or below the origin it is. The willow tree is 2 units left of the origin and 4 units above it, so its coordinates are (-2, 4).

MA.A.1.3.1

15 *Which of the following could also be written as* **three million four hundred thousand five?**

F. 3,405,000

G. 3,040,005

H. 3,045,000

I. 3,400,005

Analysis: *Choice I is correct.* Use the chart below to help read the numbers in the answer choices. When reading any number always start with the furthest left non-zero digit. It also helps to consider the digits in sets of three "between the commas". For example, in this case, the furthest left digit is a 3 in the millions place, so this part is read as three million. The next digit is a 4 in the hundred thousands place, followed by a 0 in the ten thousands place, and a 0 in the thousands place, so this part is read as four hundred thousand. The remaining digit is a 5 in the ones place and is read as five. Putting these three parts together gives us three million four hundred thousand five.

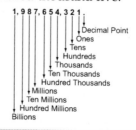

Name __Estefan bebee__

1 Which of the following shows these prime factors written in exponential form?

$2 \times 2 \times 2 \times 2 \times 5 = ?$

Ⓐ $4 \times 2^1 \times 5$

Ⓑ $4^1 \times 2 \times 5$

Ⓒ $2^4 + 5$

Ⓓ $2^4 \times 5$

2 Which of the following is the prime factorization of 84?

Ⓕ $2 \times 3 \times 7$

Ⓖ $2^2 \times 3 \times 7$

Ⓗ 2×42

Ⓘ $2 \times 6 \times 7$

3 This summer, Jed is going to spend more than a month at his aunt's house. The number of days he'll stay is a prime number. Which number could it be?

Ⓐ 23

Ⓑ 33

Ⓒ 41

Ⓓ 51

4 About $\frac{11}{100}$ of the Earth's surface water is in swamps. What is $\frac{11}{100}$ expressed in decimal form?

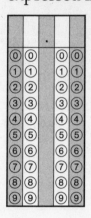

5 Mr. Jin's class has $89 left in its class activity fund at the end of the year. The class contains 22 students. If he divides the money evenly, how many dollars should he return to each student's family?

THINK
SOLVE
EXPLAIN

If there is a remainder, explain what it means.

_____

_____

_____

_____

_____

_____

Common Factors and Greatest Common Factor

Find the GCF of each pair of numbers.

1. 15, 50 ___5___   2. 6, 27 ___3___   3. 10, 25 ___5___

4. 18, 32 ___6___   5. 7, 28 ___7___   6. 54, 108 _____

7. 25, 55 _____   8. 14, 48 _____   9. 81, 135 _____

10. **Number Sense** Can the GCF of 16 and 42 be less than 16? Explain.

_____

3'4/00.005

_____

11. A restaurant received a shipment of 42 gallons of orange
juice and 18 gallons of cranberry juice. The juice needs
to be poured into equal-sized containers. What is the
largest amount of juice that each container can hold
of each kind of juice?   _____

12. At a day camp, there are 56 girls and 42 boys. The
campers need to be split into equal groups. Each has
either all girls or all boys. What is the greatest number
of campers each group can have?   _____

13. Which is the GCF of 24 and 64?

Ⓐ 4   Ⓑ 8   Ⓒ 14   Ⓓ 12

14. Do all even numbers have 2 as a factor? Explain.

THINK
SOLVE
EXPLAIN

_____

_____

MA.E.3.3.1

16 *Roosevelt conducted a survey of people's favorite beverages on two separate nights at the two most popular restaurants in Huckleberry. The results of his survey are shown in the table below. Which of the following can you determine from the information in the table?*

Beverage	Restaurant	
	Spumoni's	Donata's
Soda Pop	43	35
Orange Juice	22	20
Lemonade	17	12
Water	24	19

A. The most popular restaurant in Huckleberry.
B. The total number of people who ordered lemonade at Donata's the night of the survey.
C. The number of customers Spumoni's served the night of the survey.
D. The percent of the total votes lemonade received.

Analysis: *Choice D is correct.* You can determine the percent of votes lemonade received because it's easy to find both the total number of people who voted for lemonade and the total number of people who participated in the survey. Choice A is incorrect since you cannot draw any conclusions about favorite restaurants when no questions were asked about this, and people do not eat only at their favorite restaurant. Choices B and C are incorrect because we don't know if every customer in each restaurant was counted or surveyed.

MA.C.1.3.1

17 *The square below contains all but which of the following?*

F. parallel lines
G. obtuse angles
H. congruent sides
I. perpendicular lines

Analysis: *Choice G is correct.* Squares have parallel lines, perpendicular lines, and congruent sides, but they also have right angles, not obtuse angles.

MA.A.1.3.4

18 *Austin was building a model car. After 30 minutes, he had $\frac{7}{8}$ of the model complete. What decimal represents how much of the model he had completed after 30 minutes?*

Analysis: *The correct answer is 0.875.* To find the decimal equivalent of a fraction, divide its numerator by its denominator:
$$7/8 = 7 \div 8 = 0.875$$

MA.D.1.3.1

19 *Samantha is taller than Miranda and Carrie is taller than Charlotte. Carrie is taller than Miranda, but not Samantha. If Miranda is taller than at least one of her friends, which of the following is the correct order of the girls from tallest to shortest?*
A. Samantha, Miranda, Carrie, Charlotte
B. Samantha, Carrie, Miranda, Charlotte
C. Carrie, Samantha, Miranda, Charlotte
D. Carrie, Samantha, Charlotte, Miranda

Analysis: *Choice B is correct.* The easiest way to solve this type of problem is to make a list or series of lists that gradually incorporate what you know. For example, the first sentence says that Samantha is taller than Miranda and Carrie is taller than Charlotte. This can be reduced to two lists from taller to shorter as:

   1. Samantha   Miranda
   2. Carrie   Charlotte

The second sentence tells us that Carrie is taller than Miranda, but not Samantha, so list 1 can be modified to put Carrie between Samantha and Miranda:

   3. Samantha   Carrie   Miranda

The third sentence says that Miranda is taller than at least one of her friends. The only friend left that could be shorter than Miranda is Charlotte, so the correct order must be:

   4. Samantha   Carrie   Miranda   Charlotte

This final order also maintains the order previously established in list 2 above.

MA.B.1.3.1

20 *Gunner has a Frisbee with a diameter of 14 inches. What is the circumference of the Frisbee to the nearest hundredth of an inch? Use 3.14 as a value for π.*

Analysis: *The correct answer is 43.96 inches.* The formulas for the circumference of a circle are:

$$C = \pi d \text{ or } C = 2 \pi r$$

Since the diameter is given, use $C = \pi d$:

$$C = 3.14 \times 14; C = 43.96 \text{ inches}$$

MA.A.5.3.1

21 *Which of the following numbers is NOT a factor of 36?*
F. 6
G. 12
H. 16
I. 18

Analysis: *Choice H is correct.* A factor of a number is one of the number's divisors. That is, it evenly divides the number with no remainder. All of the choices above are divisors and, therefore, factors of 36 except Choice H, because 16 will not evenly divide 36:

$$36 \div 16 = 2.25$$

MA.A.4.3.1

22 *Kit's dog Oliver likes to dig holes. This week, Oliver dug 19 holes in the yard. ESTIMATE the number of holes Oliver will dig in the yard over a year if he digs holes at about the same rate per week all year?*
A. 500 holes
B. 1,000 holes
C. 1,200 holes
D. 1,500 holes

Analysis: *Choice B is correct.* Round off 19 holes per week to 20 and round off 52 weeks per year to 50, then multiply: 20 holes per week × 50 weeks per year ≈ 1,000 holes per year.

MA.A.1.3.2

23 *Which of the following shows the numbers in order from least to greatest?*

F. $33\%, 0.35, \dfrac{3}{5}, \dfrac{2}{3}$

G. $\dfrac{2}{3}, 33\%, 0.35, \dfrac{3}{5}$

H. $\dfrac{2}{3}, \dfrac{3}{5}, 33\%, 0.35,$

I. $\dfrac{3}{5}, 0.35, 33\%, \dfrac{2}{3}$

Analysis: *Choice F is correct.* First change all of the numbers to decimals:

$33\% = 0.33$; 0.35 is already a decimal; $\dfrac{3}{5} = 0.60$; and $\dfrac{2}{3} \approx 0.67$

The order of the decimals from least to greatest is:

$0.33, 0.35, 0.60, 0.67$ or $33\%, 0.35, \dfrac{3}{5}, \dfrac{2}{3}$, Choice F.

MA.B.1.3.2

24 *What is the measure of the angle that is* ***supplementary*** *to Angle BAC in the diagram below? Your answer will be in degrees.*

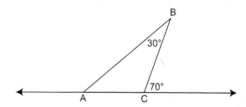

Analysis: *The correct answer is 140°.* Angle BCA measures 110° because it is supplementary to the 70° angle in the diagram, and supplementary angles always add up to 180°:

$$110° + 70° = 180°$$

Angle BAC measures 40° because the sum of the internal angles of a triangle is always 180°:

$$110° + 30° + 40° = 180°$$

The angle that is supplementary to Angle BAC measures 140° because supplementary angles always add up to 180°:

$$140° + 40° = 180°$$

MA.B.3.3.1

25 *Gunter's mom is making mashed potatoes for Thanksgiving dinner. She asks Gunter to go to the store to get her 10 pounds of potatoes. At the store, Gunter weighs one potato and sees it weighs 4.3 ounces. If each potato weighs about the same amount, ESTIMATE the number of potatoes Gunter needs to buy.*

A. 15 potatoes
B. 25 potatoes
C. 32 potatoes
D. 40 potatoes

Analysis: *Choice D is correct.* Round off the weight of each potato to approximately 4 ounces. Since there are 16 ounces to 1 pound, it takes about 4 potatoes to make a pound:

$$4 \text{ potatoes} \times 4 \text{ ounces} \approx 16 \text{ ounces of potatoes} \approx$$
$$1 \text{ pound of potatoes}$$

Gunter's mom wants 10 pounds of potatoes, so Gunter needs to buy about 40 of them:

$$4 \text{ potatoes per pound} \times 10 \text{ pounds} \approx 40 \text{ potatoes}$$

MA.E.3.3.2

26 *Renae is playing a board game with some of her friends. On her next turn, she needs to roll a 3 with a standard die to win the game. The three people before her all rolled 3s on their turns. Which of the following statements about Renae's next turn is true?*

F. Renae will probably roll a 3 because all of her friends have been able to do so.
G. Renae will probably roll a 3 because she needs one.
H. Renae will probably not roll a 3 because so many 3s have already been rolled.
I. Renae has an equal chance of rolling any number on the die.

Analysis: *Choice I is correct.* If the die is honest, then any of the six sides is equally likely to turn up. Each roll of the die is an independent event, so the probability of any particular side turning up during each roll is not affected by any past or future roll of the die.

MA.E.1.3.2

27 *The results of a pie-eating contest are shown in the table below. What was the median number of pies eaten during the contest?*

Name	Pies Eaten
Heidi	4
John	7
Summer	4
Carlos	6
Jennifer	4
J.J.	8
Monica	7

A. 4
B. 5
C. 6
D. 7

Analysis: *Choice C is correct.* The median of a data set with an odd number of values, is the middle number of the set when the values are arranged in either increasing or decreasing order. The values from this table arranged in increasing order are:

$$4, 4, 4, 6, 7, 7, 8$$

Since 6 is the middle term of these seven values, it is the median.

MA.A.3.3.1

28 *Which of the following will result in the greatest number?*

F. squaring $\dfrac{2}{5}$

G. dividing $\dfrac{2}{5}$ by 2

H. subtracting $\dfrac{1}{10}$ from $\dfrac{2}{5}$

I. multiplying $\dfrac{2}{5}$ by $\dfrac{1}{10}$

Analysis: *Choice H is correct.* Perform each operation and convert the fractional answers to decimals to see which is the largest.

For Choice F,

$$\left(\dfrac{2}{5}\right)^2 = \dfrac{2}{5} \times \dfrac{2}{5} = \dfrac{4}{25} = 0.16$$

For Choice G,

$$\dfrac{2}{5} \div 2 = \dfrac{2}{5} \times \dfrac{1}{2} = \dfrac{2}{10} = 0.20$$

For Choice H,

$$\dfrac{2}{5} - \dfrac{1}{10} = \dfrac{2}{5} \times \dfrac{2}{2} - \dfrac{1}{10} = \dfrac{4}{10} - \dfrac{1}{10}$$

$$\dfrac{3}{10} = 0.30$$

For Choice I,

$$\dfrac{2}{5} \times \dfrac{1}{10} = \dfrac{2}{50} = 0.04$$

The value for the operation in Choice H, 0.30, is larger than any of the other values, so Choice H is correct.

MA.A.1.3.3

29 *Apu baked some cookies in the shape of a heart and others in the shape of an arrow for a Valentine's Day party. Most of the cookies he baked were heart-shaped, but 40% were shaped like arrows. Which of the cookie trays below best represents this?*

A.

B.

C.

D.

Analysis: *Choice A is correct.* Each of the trays has 15 cookies. The correct tray must have 40% of the cookies arrow-shaped, so find 40% of 15:

   40% of 15 = 0.40 x 15 = 6 arrow-shaped cookies

The only tray with 6 arrow-shaped cookies is Choice A.

MA.D.2.3.1

30 *At an apple orchard, a farmer charges $10.00 to pick the first bushel of apples and $3.00 for each bushel after that. Which expression could be used to find the price of picking n bushels of apples?*

F. $(n-1) \times (\$3.00 + \$10.00)$
G. $(n-1) \times \$3.00 + \$10.00$
H. $(n-1) \times \$3.00 + \$10.00 \times n$
I. $n-1 \times \$3.00 + \$10.00$

Analysis: *Choice G is correct.* The first bushel cost $10.00 to pick so that part of the expression is represented by "+ $10.00." The rest of the bushels cost $3.00 each to pick. If there are n bushels total, then $(n-1)$ of them are picked at $3.00 each, since the other bushel was already paid for at $10.00. The expression should be some variation of $(n-1) \times \$3.00 + \$10.00$. Remember that there is usually more than one correct expression to represent a situation. Order of operations will ensure that the subtraction inside the parentheses is done first, followed by the multiplication of $3.00, and finally the addition of $10.00.

MA.A.2.3.1

31 *What is the value of the expression given below?*

$$3^3 + 3^5 - 2^5$$

Analysis: *The correct answer is 238.*

$3^3 = 3 \times 3 \times 3 = 27,\ 3^5 = 3 \times 3 \times 3 \times 3 \times 3 = 243,$ and

$$2^5 = 2 \times 2 \times 2 \times 2 \times 2 = 32,$$

so

$$3^3 + 3^5 - 2^5 = 27 + 243 - 32 = 270 - 32 = 238$$

MA.D.1.3.1

32 *What is the value of y when x = 2 in the function table below?*

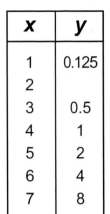

x	y
1	0.125
2	
3	0.5
4	1
5	2
6	4
7	8

Analysis: *The correct answer is 0.25 or 1/4.* The pattern is that for every increase in x by 1, y doubles. For example:

$2 \times 0.125 = 0.25;\ 2 \times 0.25 = 0.5;\ 2 \times 0.5 = 1;\ 2 \times 1 = 2;$
$2 \times 2 = 4;\ 2 \times 4 = 8$

MA.C.2.3.2

33 *Which of the following figures could be used to create a tessellation?*

A.

B.

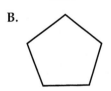

C.

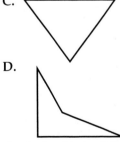

D.

Analysis: *Choice C is correct.* Remember that a tessellation is a tiling, a pattern formed by placing congruent figures together with no gaps or overlaps. All of these figures except the triangle in Choice C will leave gaps or create overlaps in an attempt to tessellate them.

MA.A.1.3.1

34 *Which of the following is the same as 0.150?*
F. fifteen tenths
G. fifteen hundredths
H. fifteen hundreds
I. fifteen thousandths

Analysis: *Choice G is correct.* The correct place names of the digits to the right of the decimal point in order are: tenths, hundredths, thousandths, ten-thousandths, and hundred-thousandths. See the diagram below. The number names that go with a given number are usually determined by the place value of the furthest right non-zero digit. Any zeros to the right of the last non-zero digit are place holders and may or may not be used. For example, the second place right of the decimal is the hundredths place, so this number can be read as fifteen-hundredths. However, you could also include the 0 in the third place right of the decimal, the thousandths place. If you do include this 0, the number is read as one hundred fifty thousandths. Both names are correct, but the shorter one is usually preferred.

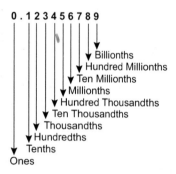

MA.B.1.3.2

35 *What is the measure of Angle CAD in the diagram below? Your answer will be in degrees.*

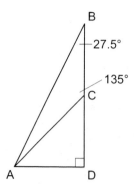

Analysis: *The correct answer is 45°.* The measure of ∠ACD is 45° because it is supplementary to ∠ACB which is given as 135° in the diagram and supplementary angles always add up to 180°:

$$135° + 45° = 180°$$

Notice that ∠CAD is in Triangle ACD along with ∠ACD and ∠ADC which is shown in the diagram as a right angle and therefore measures 90°. Since all triangles have an internal angle sum of 180°, ∠CAD measures 45°:

$$m∠CAD + m∠ACD + m∠ADC = 180°;$$
$$45° + 45° + 90° = 180°$$

MA.B.2.3.2

36 *Each year, the town of Bakersville holds the Great Cake Race. Each participant runs 2.5 miles while balancing a cake on his or her head. How long is the Great Cake Race in yards?*
A. 3,520 yards
B. 4,400 yards
C. 10,560 yards
D. 12,200 yards

Analysis: *Choice B is correct.* There are 5,280 feet in a mile and also 3 feet in a yard, so divide 5,280 by 3 to find out how many yards in a mile:

$$5,280 ÷ 3 = 1,760 \text{ yards in a mile}$$

The race is 2.5 miles long, so multiply 1,760 by 2.5:

$$2.5 \text{ x } 1,760 = 4,400 \text{ yards}$$

MA.D.2.3.2

37 *Mr. Bohr's class is conducting a science experiment and they need to convert Fahrenheit temperatures into Celsius. Use the equation below to convert 61° Fahrenheit into degrees Celsius. Round your answer to the nearest tenth.*

$$°C = \frac{5}{9} \times (°F - 32)$$

Analysis: *The correct answer is 16.1° C.* Substitute 61 into the given equation for °F;

$$°C = \frac{5}{9} \times (°F - 32); \quad °C = \frac{5}{9} \times (61 - 32);$$

$$°C = \frac{5}{9} \times (29); \quad °C = \frac{5}{9} \times \frac{29}{1}; \quad °C = \frac{145}{9};$$

$$°C \approx 16.1$$

MA.E.1.3.2

38 *The table below shows how many laps around the gym 7 different boys could run in 15 minutes. Who ran the same number of laps as the mean number of laps run?*

Name	Laps
Keith	12
Chris	7
Eddie	11
Howard	3
Archie	9
Les	10
Vic	11

F. Archie
G. Eddie
H. Howard
I. Les

Analysis: *Choice F is correct.* The mean of a data set is found by adding all the numbers in the set and dividing that sum by the number of numbers in the set. In this case,

12 + 7 + 11 + 3 + 9 + 10 + 11 = 63; 63 ÷ 7 = 9

The mean number of laps run was 9, the same as the number of laps Archie ran.

MA.E.2.3.2

39 *Heather is eating a box of animal crackers. In the box, there are 5 lions, 4 monkeys, 2 elephants, and 1 bear. If Heather takes an animal cracker from the box without looking, what are the odds she will choose an elephant?*

A. $\dfrac{2}{9}$

B. $\dfrac{1}{5}$

C. $\dfrac{1}{6}$

D. $\dfrac{1}{12}$

Analysis: *Choice B is correct.* Odds are not the same as probability. The odds for an event is defined as the ratio of the number of favorable outcomes to the number of unfavorable outcomes (in this case, number of elephant animal crackers : number of non-elephant animal crackers). There are 2 elephants and 10 crackers of other animals in the box so the odds for choosing an elephant at random are 2:10 which reduces to 1:5 and can also be written as $\dfrac{1}{5}$.

MA.A.3.3.3

40 *Curtis is planning to buy a new radio that costs $29.99, but he has a coupon for $5.00 off the lowest price of any item. When he gets to the store, the radio is on sale, marked down 10%. If sales tax is 6%, what is the total price of the radio in dollars to the nearest cent?*

Analysis: *The correct answer is $23.31.* The coupon applies to the lowest price of any item, so first find the price of the radio after the 10% discount. A 10% discount means that the new price is 90% of the regular price (100% − 10% = 90%):

90% of $29.99 = 0.90 x $29.99 = $26.99

Now deduct the $5 coupon:

$26.99 − $5.00 = $21.99

To find the total cost of the radio with sales tax, you could find 6% of $21.99 and add that amount to $21.99 or take a shortcut and find 106% of $21.99:

106% of $21.99 = 1.06 x $21.99 ≈ $23.31

MA.A.3.3.1

41 *Which of the following is an equivalent way to write* $7 \div 4$?

F. $\dfrac{1}{7} \times 4$

G. $\dfrac{1}{7} \div 4$

H. $\dfrac{7}{1} \times \dfrac{4}{1}$

I. $\dfrac{7}{1} \times \dfrac{1}{4}$

Analysis: *Choice I is correct.* Dividing by a number or multiplying by its reciprocal are equivalent operations.

$$7 \div 4 = \frac{7}{4} \text{ or } 1\frac{3}{4} \text{ or } 1.75$$

Choice I is the only operation of the four that will produce this result.

MA.B.2.3.1

42 *To the nearest degree, what is the measure of* $\angle BEC$?

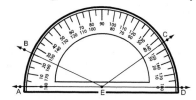

A. 25°
B. 85°
C. 120°
D. 145°

Analysis: *Choice C is correct.* Getting the correct answer requires noticing that there are two scales on the protractor and the ability to recognize the approximate size of the most common angles. For example, if you notice that $\angle BEC$ is larger than a right angle, there are two choices that makes sense, Choices C and D. However, Choice D can be eliminated because than is the measure of $\angle AEC$ and $\angle BEC$ must be smaller than this. If you are using the protractor, it doesn't matter whether you use the inside or the outside scale as long as you use the same one for both measures. For example, suppose you choose the inside scale. Notice the location of line segment EC. It is one leg of $\angle BEC$ and it is on the 35° mark. Now notice the location of line segment EB, again using the inside scale. It is the other leg of $\angle BEC$ and it is on the 155° mark. This means that the measure of $\angle BEC$ can be found by subtracting these values:

$$155° - 35° = 120°$$

MA.A.4.3.1

43 *Which of the following is the best estimate for the area of a circle with a radius of 11 inches?*

F. 200 square inches
G. 300 square inches
H. 400 square inches
I. 450 square inches

Analysis: *Choice H is correct.* The formula for the area of a circle is $\pi \times$ the radius squared. Use '3' as an estimate of pi, and calculate 11^2:

11^2; $11 \times 11 = 121$, $3 \times 121 = 363$ square inches
Recognize that the actual figure will be higher (nearer to 380) if 3.1415 is used for π, so the best estimate given is 400.

MA.E.3.3.2

44 *Rodrigo wants to survey public attitudes about Florida's rules and regulations concerning the classification of endangered, threatened, and special-concern wildlife species.* Of the sample populations listed below, which would be most likely to represent a variety of citizens' opinions?*

A. Sierra Club and World Wildlife Federation members
B. representatives of the state's timber and fishing industries
C. wildlife biologists from all of the state's public and private universities
D. a random sample from a downtown Miami street corner

*Source: The Endangered Species Protection Act; Florida Fish and Wildlife Conservation Commission

Analysis: *Choice D is correct.* Choices A and C are incorrect because these populations are much more likely to favor very strong protection and conservation efforts. Choice B is incorrect because this population is likely to favor placing economic and job issues as a higher priority. All of the above choices have some problems, but a random street corner sample is likely to be the least biased.

MA.E.1.3.1

45 *The table below lists the areas of the six largest countries in the world.*

Largest Countries in Square Miles, 2004

1	Russia	6.592.735
2	Canada	3,855,081
3	United States	3,717,792
4	China	3,705,386
5	Brazil	3,286,470
6	Australia	2,967,893

Source: www.infoplease.com

What graph best represents the data in the table?

F.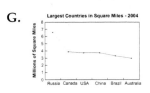

G.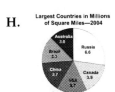

H.

I.

Analysis: *Choice F is correct.* Bar graphs are the best way to display the data in this chart because they are very good for comparing things or showing ranked items. Line graphs are usually used to show changes over time and this data is not about change; Choice G is incorrect. Circle graphs are used to show fractions of a whole; Choice H is incorrect. Some of the data is displayed incorrectly in Choice I. Canada is shown as having an area of 5 million square miles when its actual area is about 3.9 million square miles. Australia is shown as having an area of not quite 2 million square miles when its actual area is about 3 million square miles. Choice I is incorrect.

MA.A.1.3.2

46 *Which of the following is located outside the region indicated on the number line below?*

A. 1.99

B. $\frac{5}{3}$

C. $\frac{7}{3}$

D. $\frac{9}{4}$

Analysis: *Choice B is correct.* The fraction 5/3 expressed as a decimal is about 1.66. The lower limit of the shaded area on the number line appears to be about 1.9, which means that 5/3 is outside the marked region.

MA.B.2.3.2

47 *What is the best estimate of the area of the figure shown in the grid below?*

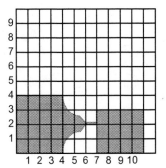

F. 20
G. 26
H. 31
I. 36

Analysis: *Choice H is correct.* There is one 4 x 4 shaded portion and one 3 x 4 shaded portion. Find these areas and add them together:

$$4 \times 4 = 16, 3 \times 4 = 12, 16 + 12 = 28$$

In addition, there are several partially-shaded squares. If parts are combined, it looks as if there are 3 or 4 more full squares, so the area of the figure must be 30-32 square units.

MA.B.2.3.1

48 Tom is building a scale model of a tall ship. Using a scale of 2 inches equals 15 feet, Tom's model is 27.3 inches long.

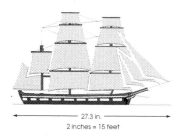

27.3 in.
2 inches = 15 feet

Determine the length of the real ship, rounded to the nearest foot.

Analysis: *The correct answer is 205 feet.* Set up and solve the following proportion:

$$\frac{2}{5} = \frac{27.3}{s} \; ; \; 2s = 15(27.3); \; 2s = 409.5;$$

$$s = 204.75 \approx 205 \text{ feet}$$

MA.B.2.3.2

49 On a map of Florida, the distance from Tallahassee to Jacksonville is about 6.25 centimeters and the distance from Jacksonville to Miami measures 13 centimeters. If the actual distance between Tallahassee and Jacksonville is 163 miles, how far is Miami from Tallahassee rounded to the nearest mile?

Tallahassee
163 miles - 6.25 cm
Jacksonville

FLORIDA

13 cm

Miami

Analysis: *The correct answer is approximately 502 miles.* The question asks how far Miami is from Tallahassee, so add the map distance from Tallahassee to Jacksonville and Jacksonville to Miami to get 19.25 cm:

$$6.25 \text{ cm} + 13 \text{ cm} = 19.25 \text{ cm}$$

Set up and solve the following proportion:

$$\frac{6.25}{163} = \frac{19.25}{d} \; ; \; 6.25d = 19.25(163); \; 6.25d = 3,137.75;$$

$$d = 502.04 \approx 502 \text{ miles.}$$

MA.C.3.3.1

50 Assume the two squares below both have sides that are 2 inches long (graphics are not drawn to scale). Square A is filled with four congruent circles that do not overlap; each circle's diameter is half the length of the side of the square. Square B contains one circle with a diameter the same length as the side of the square. The formula for the area of a circle is $A = \pi r^2$

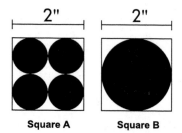

2"   2"

Square A   Square B

Which square has a greater percent of white area?

A. Square A has a greater percentage of white area.
B. Square B has a greater percentage of white area.
C. The percentage of white area is the same for both squares.
D. The percentage of white area cannot be found with the information given.

Analysis: *Choice C is correct.* The area of each square is

$$2 \text{ inches} \times 2 \text{ inches} = 4 \text{ inches}^2.$$

The area of one of the circles in Square A is πr^2, or,

$$\pi \times (0.5 \text{ inch})^2 = 0.25 \text{ inches}^2.$$

The total area of the four circles is:

$$4 \times 0.25\pi \text{ inches}^2 = 1\pi \text{ inches}^2 = \pi \text{ inches}^2.$$

The area of the circle in Square B is:

$$\pi(1 \text{ inch})^2 = 1\pi \text{ inches}^2 = \pi \text{ inches}^2.$$

Mathematics Assessment One—Correlation Chart

The Correlation Charts can be used by teachers to identify areas of improvement. When students miss a question, place an "X" in the corresponding box. A column with a large number of "Xs" shows more practice is needed with that particular standard.

Correlation	MA.B.2.3.2	MA.A.3.3.3	MA.E.3.3.1	MA.A.3.3.2	MA.C.2.3.1	MA.C.1.3.1	MA.D.1.3.2	MA.D.2.3.2	MA.E.1.3.3	MA.E.2.3.2	MA.A.3.3.3	MA.E.2.3.1	MA.B.1.3.3	MA.C.3.3.2	MA.A.1.3.1	MA.E.3.3.1	MA.C.1.3.1	MA.A.1.3.4	MA.D.1.3.1	MA.B.1.3.1
Answer	B	G	B	I	C	1,080	F	D	F	D	H	A	G	C	I	D	G	0.875	B	43.96 inches
Question	1	2	3	4	5	6	7	8	9	10	11	12	13	14	15	16	17	18	19	20

Student Names

Mathematics Assessment One—Correlation Chart

Correlation	MA.A.5.3.1	MA.A.4.3.1	MA.A.1.3.2	MA.B.1.3.2	MA.B.3.3.1	MA.E.3.3.2	MA.E.1.3.2	MA.A.3.3.1	MA.A.1.3.3	MA.D.2.3.1	MA.A.2.3.1	MA.D.1.3.1	MA.C.2.3.2	MA.A.1.3.1	MA.B.1.3.2	MA.B.2.3.2	MA.D.2.3.2	MA.E.1.3.2	MA.E.2.3.2	MA.A.3.3.?
Answer	H	B	F	140	D	I	C	H	A	G	238	0.25	C	G	45	B	16.1°C	F	B	$23.31
Question	21	22	23	24	25	26	27	28	29	30	31	32	33	34	35	36	37	38	39	40

Student Names

Mathematics Assessment One—Correlation Chart

Correlation	MA.A.3.3.1	MA.B.2.3.1	MA.A.4.3.1	MA.E.3.3.2	MA.E.1.3.1	MA.A.1.3.2	MA.B.2.3.2	MA.B.2.3.1	MA.B.2.3.2	MA.C.3.3.1
Answer	I	C	H	D	F	B	H	2003 FCAT reading	2012 reading	C
Question	41	42	43	44	45	46	47	48	49	50

Student Names

Mathematics Assessment Two Introduction Page as it appears on page 139 in the
Show What You Know® on the 7th Grade FCAT, Mathematics Student Workbook.

Mathematics Assessment Two

Directions for Taking the Mathematics Assessment Two

On this section of the Florida Comprehensive Assessment Test (FCAT), you will answer 50 questions.

For multiple-choice questions, you will be asked to pick the best answer out of four possible choices and fill in the answer in the answer bubble. On gridded-response questions, you will also fill your answer in answer bubbles, but you will also fill in numbers and symbols corresponding to the solution you obtain for a question. Fill in the answer bubbles and gridded-response answer bubbles on the Answer Sheets on pages 163–167 to mark your selection.

Read each question carefully and answer it to the best of your ability. If you do not know an answer, you may skip the question and come back to it later.

Figures and diagrams with given lengths and/or dimensions are not drawn to scale. Angle measures should be assumed to be accurate. Use the Mathematics Reference Sheet on page 31 to help you answer the questions.

When you finish, check your answers.

MA.A.1.3.1

1 *Which of the following numbers is equal to seven million three hundred five thousand four hundred one?*

A. 7,000,305,401
B. 7,350,401
C. 7,305,410
D. 7,305,401

Analysis: *Choice D is correct.* Use the chart below to help read the numbers in the answer choices. When reading any number always start with the furthest left non-zero digit. It also helps to consider the digits in sets of three "between the commas". For example, in this case, the furthest left digit is a 7 in the millions place for Choices B, C, and D, but in the billions place for Choice A, so Choice A can be eliminated. The written name in the question for the next three digits is "three hundred five thousand", so the next three digits should be a 3 in the hundred thousands place, followed by a 0 in the ten thousands place, and a 5 in the thousands place. This eliminates Choice B. The remaining part of the number is read as "four hundred one," so the next three digits should be a 4 in the hundreds place, followed by a 0 in the tens place, and a 1 in the ones place. Only Choice D fits all of these requirements.

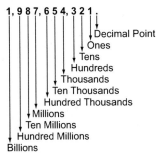

MA.A.1.3.3

2 *A real estate agent will receive a 3% commission from the sale of a house that sold for $218,800. How much money will the agent earn from this sale?*

F. $6,024
G. $6,564
H. $7,800
I. $12,200

Analysis: *Choice G is correct.* To find the agent's commission, multiply the sale amount by 3%, or 0.03:
$218,800 x .03 = $6,564

MA.A.4.3.1

3 *Which of the following would be an appropriate situation in which to use an exact figure rather than an estimate?*

A. Cutting a piece of string to wrap a bouquet of flowers
B. Deciding how much paper will be needed to wrap a gift
C. Planning how much time is needed to prepare a cake for a party
D. Figuring how much change you should receive from the dollar you gave a clerk when purchasing a pack of gum

Analysis: *Choice D is correct.* Estimates work fine in all of these situations, except making change. Exact figures tend to work best when there is some consequence for not having an exact amount, such as making change or cutting an expensive picture frame too small to fit the picture it is being built for.

MA.B.1.3.3

4 *A small circular parachute has a radius of 20 feet. The larger model has a radius that is 40% bigger. Which of the following best explains why the area of the larger parachute is not 40% larger than the smaller parachute?*

F. The radius is not used for computing the area.
G. Since the radius is squared in the area formula, the 40% increase in radius size leads to a 96% increase in area.
H. Since the radius is by definition that of the diameter, it must be doubled before it can be used in place of diameter in area computations.
I. This can be explained by the formula for the area of a circle and the tendency people have to round pi (π) to 3.14, instead of using a precise value.

Analysis: *Choice G is correct.* The area of a circle is found with the formula $A = \pi r^2$. The small parachute has a radius of 20 feet, so its area is:
$$A = \pi(20)^2; A = 400\pi \text{ square feet.}$$
The larger parachute has a radius 40% larger, so its radius is 28 feet:
$$(1.4 \times 20 = 28)$$
The area of the larger parachute is:
$$A = \pi(28)2; A = 784\pi \text{ square feet.}$$
The increase in area is 384π square feet:
$$(784\pi - 400\pi = 384\pi)$$
So the percentage increase in area is 96%:
$$(384\pi \div 400\pi = .96 = 96\%)$$

MA.B.1.3.1

5 *A rectangular prism with a width of 20 inches and a length of 60 inches has a volume of 2,400 cubic inches. What is the height of this prism? (Note: Prism in diagram not drawn to scale.)*

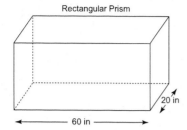

Rectangular Prism

20 in

60 in

A. 2 inches
B. 10 inches
C. 12 inches
D. 24 inches

Analysis: *Choice A is correct.* To find the height of this prism, use the formula for volume of a prism, $V = l \times w \times h$, and using $l = 60$, $w = 20$, and $V = 2400$. Since $20 \times 60 \times h = 2,400$, we know that $1,200 \times h = 2,400$. Therefore, $h = 2,400 \div 1,200$, or 2 inches.

MA.E.1.3.2

6 *George is taking a class in which the students are given six tests, and the mean of the test grades determines a student's final grade. George's grades so far are 88, 76, 82, 90, and 70. Which of the following best describes what will happen to his grade if George skips the last test and receives a zero on it?*

F. Since George already has a good run of passing test scores, it will have little to no effect on his grade.
G. George's grade will drop from a solid "B" to a solid "C."
H. George will be in danger of failing the class, unless 68 is considered a passing grade at his school.
I. Cannot be determined from the information given.

Analysis: *Choice H is correct.* George's current average is 81.2:

$$88 + 76 + 82 + 90 + 70 = 406; 406 \div 5 = 81.2$$

A zero would plunge his average into the 60s:

$$88 + 76 + 82 + 90 + 70 + 0 = 406;$$
$$406 \div 6 \approx 67.7$$

MA.D.1.3.2

7 *Look at the graph below.*

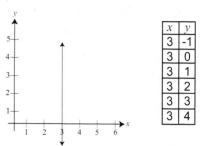

x	y
3	-1
3	0
3	1
3	2
3	3
3	4

Which of the following expressions correctly describes the line shown?

A. $y = 3$
B. $3x = 1$
C. $x = 3$
D. $3y = 1$

Analysis: *Choice C is correct.* A vertical line has an equation where x is constant and a y term is not even a part of the equation. For example, $x = 3$ is the equation of a vertical line (parallel to the y-axis) passing through $(3, 0)$, $(3, 1)$, $(3, 2)$, $(3, 3)$, etc. In the equation $x = 3$, it doesn't matter what y is, because y is not a part of the equation, but x is always 3. Another way to think about this is to consider the slope-intercept form for the equation of a line, $y = mx + b$. In this form, the m is the slope of the line and the b is the y-intercept (the place where the line crosses the y-axis). An equation like $x = 3$ can never be put into slope-intercept form since there is no y to put on the left side of the equation, and also because, 1. it has no slope since slope is undefined for a vertical line and 2. It has no y-intercept because a vertical line will never cross the y-axis.

MA.A.3.3.1

8 *Which of the following symbols would create a true mathematical statement when placed between these two expressions?*

$$-5 \times -5 \boxed{} - (5 \times 5)$$

F. $=$
G. $<$
H. $\leq$
I. $>$

Analysis: *Choice I is correct.* The expression -5×-5 means to multiply -5 by -5, i.e. $-5 \times -5 = 25$. The expression $-(5 \times 5)$ means to multiply 5 by 5 and then make the result negative, i.e. $-(5 \times 5) = -(25) = -25$. Since $25 > -25$, Choice I is correct.

MA.A.3.3.2

9 *The table below lists some interesting geography facts about the earth.*

Earth Geography Facts	
Age (Approximate)	4.55 billion years
Total Area:	196.940 million square miles
Land Area:	57.506 million square miles
Water Area:	139.434 million square miles
Terrain: Highest Land Elevation: Mt. Everest	29,035 feet
Lowest Land Elevation: Dead Sea	− 1,349 feet below sea level
Greatest Ocean Depth: Mariana Trench	− 35,840 feet below sea level
Land Use: Arable Land (Farmable)	13.31%

*The World Factbook, 2006; https://www.cia.gov/cia/publications/factbook/geos/xx.html

Use the information in the table above to calculate the percent of the earth's surface that is covered by water. Round your answer to the nearest tenth of a percent.

Analysis: *The correct answer is 70.8%.* In order to calculate the percent of Earth's surface covered by water, divide the water area of the earth by its total area and change the decimal into a percent

$$(139.434 \div 196.940 \approx 0.708 \approx 70.8\ \%)$$

All the other information in the table above is irrelevant (unnecessary) to this question.

MA.E.3.3.2

10 *A campground is asking weekend users of the facility to fill out a short survey. The campground owners want to learn more about their customer base in order to provide better services. Which of the following questions is too biased to be of much use in the survey?*

A. How many days per year, on average, do you spend camping?

B. Do you think 10 PM through 6 AM is long enough for quiet hours?

C. Why do you think the other campground down the road is so run-down looking?

D. Was there an adequate amount of hot water in the bathhouse during your entire time here?

Analysis: *Choice C is correct.* The question in Choice C draws a conclusion by stating that "the other campground down the road is so run-down looking," instead of simply collecting information. It inserts the opinion of the survey writer, which shows bias toward one campground.

MA.E.1.3.1

11 *The table below lists the eight largest cities in Florida.*

Population of the Largest Cities in Florida		
1	Jacksonville	782.623
2	Miami	386.417
3	Tampa	325.989
4	St. Petersburg	249.079
5	Hialeah	220.485
6	Orlando	213.223
7	Fort Lauderdale	167.380
8	Tallahassee	158.500
	(2005 estimate)	

Which graph best represent the data in the table?

F.
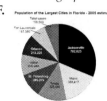

G.

H.
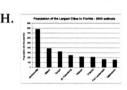

I.
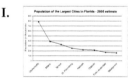

Analysis: *Choice H is correct.* A bar graph is the best way to display the data in this chart because bar graphs are very good for comparing things or showing ranked items. Choice F is incorrect because circle graphs are used to show fractions of a whole. Choice G is incorrect because scatter plots are used to discover trends or correlation between two sets of values. Choice I is incorrect because line graphs are usually used to show changes over time and this data is not about change.

MA.A.3.3.3

12 *The first Egyptian pyramid, the famous Step Pyramid of Saqqara, was built for the burial of Pharaoh Djoser by his Vizier in about 2800 B.C.E. This pyramid still stands. Approximately how old is this pyramid to the nearest 100 years.*

Analysis: *The correct answer is about 4800 years.* The most common way of calculating the passage of time is to subtract the earlier date from the later date. For example, to calculate how old the United States is, subtract the year of the Declaration of Independence from the current year, for example:

$$2006 - 1776 = 230 \text{ years.}$$

To use this method with B.C.E. dates consider them as negative numbers, for example, to calculate the age of the Step Pyramid:

$$2006 - (-2800) = 4806 \approx 4800 \text{ years old.}$$

The approximate symbol, $\approx$, is appropriate here because the scientists have had to guess when the pyramid was started and when it was finished.

MA.D.1.3.1

13 *Which of the following patterns behaves according to a* $\times$ *2, a + 2 alternating rule?*

A. 2, 4, 8, 10, 16, 18
B. 1, 4, 7, 14, 24, 26
C. 1, 3, 5, 9, 11, 15
D. 1, 2, 4, 8, 10, 20

Analysis: *Choice D is correct.* The number 2 is twice the number 1, and 4 is equal to 2 + 2. Continuing through the list, the number 8 is twice the number 4, and 10 is equal to 8 + 2. Finally, the number 20 is twice the number 10.

MA.A.2.3.1

14 *Stars and galaxies are so far away that astronomers do not measure their distance in miles. Distances like these are measured in light years. A light year is the distance that light can travel in a year and light travels very rapidly indeed-approximately 186,282 miles per second. To find out how far this is in miles, multiply: 186,282 miles/sec* $\times$ *60 sec/min* $\times$ *60 min/hr* $\times$ *24 hr/day* $\times$ *365 days/year. When you find this product on a calculator you get a number in scientific notation: about 5.87* $\times$ *1012. How many miles is this in standard notation?*

F. 5,871,012
G. 587,000,000,000 miles
H. 5,870,000,000,000 miles
I. 587,000,000,000,000 miles

Analysis: *Choice H is correct.* A number in scientific notation has two parts: a decimal number between 1 and 10, and a power of 10. When converting a number in scientific notation to standard notation, the exponent on the power of 10 tells which way to move the decimal and how many places to move it. If the exponent is negative, move the decimal to the left. If the exponent is positive, move the decimal to the right. This exponent is a positive 12, so the decimal must move 12 places to the right. Since there are only two digits, an eight and a seven, to the right of the decimal, add 10 zeros after the seven to obtain 5870000000000, then put in commas every three digits:

$$5{,}870{,}000{,}000{,}000$$

MA.A.1.3.4

15 *Which of these choices does not belong with the other three*

A. $\dfrac{21}{3}$

B. $\left| -7 \right|$

C. 7°

D. $\sqrt{49}$

Analysis: *Choice C is correct.* All of the other choices are equivalent to 7, but Choice C is equivalent to 1 because any number (except 0) raised to the zeroth power is 1.

MA.B.2.3.1

16 *Look carefully at the drawing below.*

Use this centimeter ruler to find the best estimate of the length of the pencil.

F.   11 cm
G.   11.3 cm
H.   11.8 cm
I.   12 cm

Analysis: *Choice H is correct.* If you look closely, you can see that the tip of the pencil is between the 11 cm mark and the 12 cm mark. That eliminates Choices F and I. It appears that the pencil's tip is closer to the 12 cm mark than the 11 cm mark. That makes Choice H a better answer than Choice G.

MA.B.2.3.2

17 *The large wooden crate below was built to ship machine parts.*

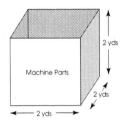

What is its volume in cubic feet?

A.   8 cubic feet
B.   24 cubic feet
C.   72 cubic feet
D.   216 cubic feet

Analysis: *Choice D is correct.* There are two ways to do this problem. You can either find the volume of the crate in cubic yards and then convert it to cubic feet, or you can convert the crate's dimensions to feet and then find the volume directly in cubic feet. Using the first method, the volume of any rectangular prism including a cube can be found with the formula: $V = lwh$. In this case, $l = 2$, $w = 2$, and $h = 2$, so the volume of this crate is 8 cubic yards:

$$(V = lwh; V = 2 \times 2 \times 2; V = 8)$$

There are 27 cubic feet in a cubic yard. This is because a cubic yard is 1 yard long, 1 yard wide, and 1 yard high. Since there are 3 feet in a yard, a cubic yard could also be measured as 3 feet long, 3 feet wide, and 3 feet high, so:

$$3 \times 3 \times 3 = 27$$

Since the crate has a volume of 8 cubic yards and there are 27 cubic feet in a cubic yard, the crate has a volume of 216 cubic feet ($8 \times 27 = 216$). Looking at the problem a different way, first convert the dimensions of the crate to feet. Since there are 3 feet to each yard, then the length, width, and height of this crate are all 6 feet ($2 \times 3 = 6$). The volume of a rectangular prism can be found with:

$$V = lwh, \text{ so } V = 6 \times 6 \times 6 = 216 \text{ cubic feet.}$$

MA.B.1.3.3

18 *Look at the diagrams below.*

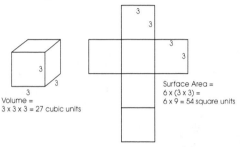

If a cube of 3 units on each edge is doubled to 6 units on each edge, how is the surface area and volume affected?

F.   Both the surface area and the volume are doubled.
G.   Both the surface area and the volume are multiplied by 4.
H.   The surface area is doubled and the volume is multiplied by 4.
I.   The surface area is multiplied by 4 and the volume is multiplied by 8.

Analysis: *Choice I is correct.* The surface area and volume of the small cube have already been calculated, so calculate the surface area and volume of the large cube:

$$\text{S.A.} = 6 \times (6 \times 6) = 6 \times 36 = 216 \text{ square units;}$$
$$V = 6 \times 6 \times 6 = 216 \text{ cubic units.}$$

Now find the ratio of the large cube's surface area to the small cube's surface area: 216/54 which reduces to 4/1. This means that when the cube's dimensions double, the surface area is multiplied by 4 which eliminates Choices F and H. Find the ratio of the large cube's volume to the small cube's volume: 216/27 which reduces to 8/1. This means that when the cube's dimensions double, the volume is multiplied by 8 which eliminates all choices except Choice I.

MA.E.3.3.1

19 *Batting average is the decimal representation of hits a batter achieves divided by the number of at-bat opportunities he or she has. A batting average of .300 means that for every 10 times at bat, the player gets 3 hits. A .300 batting average is very good. A professional baseball player has a 10-year batting average of .315 through the previous season. This year his average is .120. Which of the following statements would be difficult to defend based on the known data?*

A. This player's career is washed up.

B. This player is in a slump.

C. This player is poised to bat .500 in the next few games.

D. This player is batting about what he always does.

Analysis: *Choice D is correct.* The data shows that the player has had an excellent batting average and may be in a slump, poised to break out, or at the end of his career. The only statement that would be difficult to defend using this data is that his present performance is typical of his usual batting ability.

MA.C.2.3.1

20 *In the drawing below, triangle HIJ is similar to triangle KLJ.*

What is the measure of segment KL?

F. 10 cm

G. 6 cm

H. 4 cm

I. 3 cm

Analysis: *Choice I is correct.* Similar triangle are the same shape, but not necessarily the same size. Two things are always true about similar triangles; first, corresponding angles of similar triangles are congruent and second, corresponding sides of similar triangles are proportional. This means that you have to compare the correct parts of each triangle to get the correct answer. For example, in this drawing, angle I and angle L are congruent, so the sides opposite each, segments HJ and KJ are corresponding and proportional. Also, segments IJ and LJ are corresponding and proportional as are segments HI and KL. Being proportional means that the same ratio describes the relationship between all pairs of corresponding sides. In this case,

$$\frac{HJ}{KJ} = \frac{HI}{KL}; \quad \frac{27}{6} = \frac{13.5}{x}$$

cross multiplying $27x = 6(13.5)$; $27x = 81$; $x = 3$.

MA.B.1.3.2

21 *What is the measure of Angle XYZ in the triangle below? Your answer will be in degrees.*

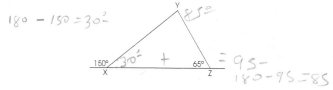

Analysis: *The correct answer is 85°.* Two facts are needed to solve this problem. First, a straight line has 180° and second, the sum of the angles inside a triangle is 180°. Using these facts, angle YXZ must be 30° since the adjacent angle is 150° and together they make up a straight line:

$$150° + 30° = 180°.$$

Two interior angles of the triangle are now known. Simply add them together and subtract their sum from 180° to find the third angle:

$$30° + 65° = 95°; 180° - 95° = 85°$$

The measure of angle XYZ is 85°.

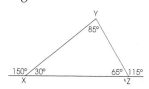

MA.C.3.3.2

22 *Look at the drawing below.*

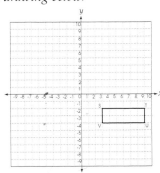

If rectangle STUV is translated left eight units and up three units, what are the new coordinates of Point V?

A. (-5, -1)
B. (1, -1)
C. (1, 1)
D. (-5, 1)

Analysis: *Choice A is correct.* When a figure is translated, every point and line segment moves the same amount in the same directions. The figure's size, shape, and orientation do not change. In this case every point moves left eight units and up three units. This is the same as subtracting eight from each point's original *x*-coordinate and adding three to each point's original *y*-coordinate. See the table and graph below. Choice B is incorrect because it's the new location of Point U, not Point V. Choice C is incorrect because it's the new location of Point T, not Point V. Choice D is incorrect because it's the new location of Point S, not Point V.

Point	Original Coordinates	Coordinates After Translation (x - 8, y +3)
S	(3, -2)	(-5, 1)
T	(9, -2)	(1, 1)
U	(9, -4)	(1, -1)
V	(3, -4)	(-5, -1)

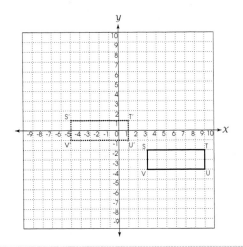

MA.C.1.3.1

23 *Which net below can be folded to create a cube?*

F.

G.

H.

I.

Analysis: *Choice I is correct.* A net is a two-dimensional representation of a three-dimensional figure. Figures F and H are incorrect because they both have four squares sharing a common point, which prevents them from being separate faces of a three-dimensional figure. Choice G is incorrect because it will form an open-ended cube. Cubes have 6 faces, but only 5 are represented in this net.

MA.D.1.3.2

24 *The table below uses a function to calculate a y-value for each value of x.*

x	y
0	-1
1	1
2	3
3	5
4	7

Which of these sentences tells how to find the correct value of y for each value of x in the table?

A. Find *y* by subtracting one from *x*.
B. Find *y* by multiplying *x* by two.
C. Find *y* by multiplying *x* by two and then subtracting one.
D. Find *y* by dividing *x* by one and then subtracting one.

Analysis: *Choice C is correct.* In order to be correct the sentence must be able to calculate every y for every x in the table.

When $x = 0$, $y = 2 \times 0 - 1 = 0 - 1 = -1$.
When $x = 1$, $y = 2 \times 1 - 1 = 2 - 1 = 1$.
When $x = 2$, $y = 2 \times 2 - 1 = 4 - 1 = 3$.
When $x = 3$, $y = 2 \times 3 - 1 = 6 - 1 = 5$.
When $x = 4$, $y = 2 \times 4 - 1 = 8 - 1 = 7$.

Choices A and D are really the same rule. Whenever you divide a number by one you get the same number. These rules only hold for the first step: When $x = 0$, $y = 0 - 1 = -1$. In order to be correct, the rule must hold for every step in the table. Choice B is incorrect because it does not calculate the correct value of y for any x in the table, not even the first one.

MA.D.2.3.1

25 *Tom wants to find a summer job that pays at least $5 per hour. Which of the choices below correctly graphs the statement "at least $5 per hour?"*

F. ![number line -8 to 8, open circle at 5 shading left]
-8 -7 -6 -5 -4 -3 -2 -1 0 1 2 3 4 5 6 7 8

G. ![number line -8 to 8, filled circle at 5 shading right]
-8 -7 -6 -5 -4 -3 -2 -1 0 1 2 3 4 5 6 7 8

H. ![number line -8 to 8, filled circle at 5 shading left]
-8 -7 -6 -5 -4 -3 -2 -1 0 1 2 3 4 5 6 7 8

I. ![number line -8 to 8, open circle at 5 shading right]
-8 -7 -6 -5 -4 -3 -2 -1 0 1 2 3 4 5 6 7 8

Analysis: *Choice G is correct.* The statement "at least $5 per hour" means that $5 per hour or more would be acceptable to Tom. This statement can be represented mathematically as $w \geq 5$ where w represents a wage acceptable to Tom. Since $5 per hour is acceptable, a filled-in circle should appear on 5 in the graph of this equation, and everything to the right of 5 should also be highlighted because those numbers also represent hourly wages acceptable to Tom. Choices F ($w < 5$) and H ($w \leq 5$) are incorrect because they represent wages of less than $5 per hour, less than or equal to $5 per hour, or even, in the case of the negative numbers, the possibility that Tom would have to pay his employer to work there. Choice I represents the case of more than $5 an hour, so the lowest acceptable wage under this condition would be $5.01.

MA.A.1.3.2

26 *Which of the following choices places the numbers in order from least to greatest?*
43⌐

A. 2^3, $\left| -9 \right|$, $\sqrt{65}$, $\dfrac{43}{5}$

B. 2^3, $\sqrt{65}$, $\dfrac{43}{5}$, $\left| -9 \right|$ = 9

C. $\left| -9 \right|$, $\dfrac{43}{5}$, $\sqrt{65}$, 2^3

D. $\left| -9 \right|$, 2^3, $\sqrt{65}$, $\dfrac{43}{5}$

Analysis: *Choice B is correct.* To make it easier to compare the expressions, convert all of them to an equivalent decimal:

$$2^3 = 2 \times 2 \times 2 = 8;$$
$\sqrt{65}$ is a little less than 8.1 because
$$8 \times 8 = 64 \text{ and } 8.1 \times 8.1 = 65.61;$$
$$\dfrac{43}{5} = 8.6;$$

$\left| -9 \right| = 9$ because the absolute value of any number is positive.

MA.A.5.3.1

27 *What is the next number in the sequence below?*

$5, 7, 11, 13, 17, 19, 23, \underline{}$ 25

Analysis: *The correct answer is 29. This is a sequence of prime numbers.*

MA.A.1.3.1

28 *Which of the following is the same as 0.000170?*
F. seventeen millionths
G. seventeen hundred thousandths
H. one hundred seventy ten thousandths
I. seventeen ten thousandths

Analysis: *Choice G is correct.* The correct place names of the digits to the right of the decimal point in order are: tenths, hundredths, thousandths, ten-thousandths, hundred-thousandths, and millionths. See the diagram below. The number names that go with a given number are usually determined by the place value of the furthest right non-zero digit. Any zeros to the right of the last non-zero digit are place holders and may or may not be used. For example, the fifth place right of the decimal is the hundred thousandths place, so this number can be read as seventeen hundred thousandths. However, you could also include the 0 in the sixth place right of the decimal, the millionths place. If you do include this 0, the number is read as one hundred seventy millionths. Both names are correct, but the shorter one is usually preferred.

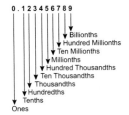

```
0.123456789
       │││││││││
       ▼ Billionths
        ▼ Hundred Millionths
       ▼ Ten Millionths
       ▼Millionths
       ▼ Hundred Thousandths
      ▼ Ten Thousandths
      ▼ Thousandths
     ▼Hundredths
    ▼ Tenths
  Ones
```

MA.B.1.3.4

29 *Ted is building a model of an antique car. The model is 22.5 inches long and the box states that it's a 1:8 scale model.*

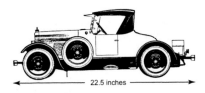

←————— 22.5 inches —————→

How long is the real car rounded to the nearest foot?
A. 8 feet
B. 12 feet
C. 14 feet
D. 15 feet

Analysis: *Choice D is correct.* This problem can be solved with a proportion:

$$\frac{1}{8} = 22.5/x$$

Cross multiply to solve for x:
$$1 \times x = 8 \times 22.5; \, x = 180 \text{ inches}$$
The car is 180 inches long, but the questioned asked for its length in feet, so divide 180 by 12:
$$180 \div 12 = 15 \text{ feet}$$

MA.B.3.3.1

30 *Shelley divided 2,400 by 3.8 and got an answer of .6315. Which of the following best describes why this answer isn't reasonable?*
F. You can't divide a whole number by a decimal and get a decimal number.
G. You should get an answer of about 60, since 3.8 must be changed to 38 by moving the decimal when the problem is set up.
H. You should get an answer of about 600, since 2,400 divided by 4 would be 600.
I. You should get an answer of about .0015, a very small decimal number.

Analysis: *Choice H is correct.* You should round 3.8 to 4 for easier division because 2,400 divided by 4 is 600, so when you divide 2,400 by 3.8 you get an answer of about 600.

MA.B.1.3.1

31 *Quality Cardboard Box Company needs a general formula to calculate the amount of cardboard they need to make any box. The diagrams show two different views of the same box whose dimensions are l, w, and h for the box's length, width, and height in inches.*

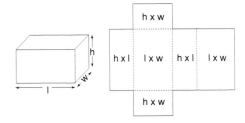

Which of these expressions correctly calculates the box's surface area, S.A.?
A. S.A. = $l \times w \times h$
B. S.A. = $(l \times w) + (h \times l) + (h \times w)$
C. S.A. = $(2 \times l \times w) + (2 \times h \times l) + (2 \times h \times w)$
D. S.A. = $6 \times l \times w$

Analysis: *Choice C is correct.* There are three different kinds of rectangles in the net of this box and there are two copies of each type.

MA.E.1.3.2

32 *There are eight houses on a particular block. All of these houses' values have been appraised at different amounts under $150,000 except for a single house that is worth $2,500,000. Which of the following would be the best measure of central tendency for representing the value of a typical house on this block?*

F. mean
G. median
H. mode
I. range

Analysis: *Choice G is correct.* Since all of the houses are appraised at different amounts under $150,000, except for a single house that is worth $2,500,000, there is no mode for this set. The single high-priced house would skew the mean, making the typical house seem more valuable than it really is. Range is not a measure of central tendency, so it can't represent typical value. Since all of the houses have different appraisals, the median, or middle number, offers the best representation.

MA.A.5.3.1

33 *Which of the following statements is NOT true about the set of numbers below?*

80, 60, 120, 420, 840

A. All of the numbers are even numbers.
B. All of the numbers are multiples of 4.
C. All of the numbers are factors of 3,360.
D. All of the numbers are multiples of 40.

Analysis: *Choice D is correct.* All of the numbers are even, multiples of 4, and factors of 3,360. In addition, all of the numbers are multiples of 40, except 60 and 420. A multiple of a number is the product of that number and an integer. There is no integer, i, such that

$$40 \times i = 60 \text{ or } 40 \times i = 420$$
$$(60 \div 40 = 1.5 \text{ and } 420 \div 40 = 10.5).$$

MA.C.1.3.1

34 *What is the value of x in the diagram below?*

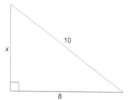

F. $\sqrt{12}$

G. $\sqrt{36}$

H. $\sqrt{100}$

I. $\sqrt{164}$

Analysis: *Choice G is correct.* Plug the values of the given sides into the Pythagorean Theorem to find the correct answer:

$$8^2 + x^2 = 10^2; \ \ 64 + x^2 = 100$$

Solve for x:

$$64 + x^2 = 100; \ x^2 = 100 - 64; \ x^2 = 36;$$

$$\sqrt{x^2} = \sqrt{36} \ \ ; x = 6.$$

MA.A.3.3.3

35 *In the year 2000, the State of Florida had 6,910,168 workers aged 16 or older. Of these, 5,445,527 drove alone to get to work while 893,766 carpooled. Approximately what percent of workers in Florida carpooled to work?*

A. about 89.4%
B. about 78.8%
C. about 16.4%
D. about 12.9 %

*Source: U.S. Census Bureau; www.census.gov/prod/2004pubs/c2kbr-33.pdf

Analysis: *Choice D is correct.* To find the percent of workers in Florida who carpooled to work, divide the number of workers who carpooled by the total number of workers and then convert the decimal to a percent:

$$893,766 \div 6,910,168 \approx 0.129340 \approx 0.129 \approx 12.9\%$$

MA.B.3.3.1

36 *In 1984, fisherman Gary Merriman caught an enormous 1,656-pound blue marlin off the coast of Kona, Hawaii. The actual fish was fifty times larger than the drawing below.*

Approximately how long was the actual fish in feet?

F. 4 feet
G. 16 feet
H. 17 feet
I. 200 feet

Analysis: *Choice H is correct.* The fish in the drawing is 4 inches long. If the actual fish is 50 times this size, it must be 200 inches long (4 x 50 = 200). The question asks for its length in feet, so divide 200 by 12:

$$(200 \div 12 \approx 16.66666667 \approx 17 \text{ feet})$$

MA.C.2.3.1

37 *Which of the following statements is true of congruent triangles?*

A. If all of the angles of one triangle are equal to all of the angles of another triangle, the triangles must be congruent.
B. If two triangles have equal corresponding angles and proportional corresponding sides, then the triangles must be congruent.
C. If three sides of one triangle are equal in measure to the three sides of another triangle, then the triangles must be congruent.
D. If two triangles have longest sides with the same measure, then the triangles must be congruent.

Analysis: *Choice C is correct.* If three sides of one triangle are equal in measure to the three sides of another triangle, then the triangles must be congruent. This statement about equal sides producing congruent figures only holds for triangles however. It does not hold for polygons with four or more sides. One way to prove that the other examples are false is to find a counterexample.

A counterexample shows one instance where the statement doesn't hold true. If you can find even one valid counterexample, then the statement is false. For example, Choice A is incorrect. See the counterexample below. Choice B is incorrect because it defines similar triangles, not congruent triangles. Choice D is also incorrect. See the counterexample below.

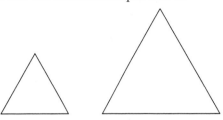

Two non-congruent triangles with all angles equal.

Two non-congruent triangles whose longest sides have the same measure.

MA.E.3.3.1

38 *In a science class of 25 people, 13 people scored between 80% and 90% on the last test. Which of the following must be true?*

F. The mean score will fall between 80% and 90%.
G. The mode of the test scores will fall between 80% and 90%.
H. The median of the test scores will fall between 80% and 90%.
I. The maximum score is closer to the mean score than the minimum score.

Analysis: *Choice H is correct.* When a set of numbers is ordered from least to greatest, the number in the middle represents the median. If 13 people scored between 80% and 90%, the middle term of the set of 25 scored between 80% and 90%.

MA.D.1.3.1

39 *Read the following statement.*

The product of a number and seven decreased by two is twelve.

Which expression correctly represents this statement?

A. $(x \div 7) - 2 = 12$
B. $7x - 2 = 12$
C. $(7 + x) - 2 = 12$
D. $7x \div 2 = 12$

Analysis: *Choice B is correct.* The word "product" indicates multiplication should be used. The term "a number" refers to some unknown number that we can represent with a variable, in this case x. "Decreased by" indicates subtraction should be used and "is" means equals. Choice A is incorrect because it represents the quotient of a number and seven, not the product of a number and seven. Choice C is incorrect because it represents the sum of a number and seven, not the product of a number and seven. Choice D is incorrect because "decreased by" indicates subtraction, not division, should be used.

MA.E.3.3.2

40 *Tim was doing an experiment for his statistics class. He flipped a coin nine times in a row and it landed on "heads" each time. He predicted that the coin was more likely to land on tails on the 10th flip. Which choice is the best evaluation of his prediction?*

F. Tim is wrong because whether the coin landed on "heads" or "tails" is independent of the results of previous tosses.
G. Tim is correct because probability states that for an even number of flips, every other flip should land on "tails."
H. Tim is wrong because probability states that for an even number of flips, every other flip should land on "heads."
I. Tim is correct, because it landed on "heads" so many times, the probability that it will land on "heads" again is zero.

Analysis: *Choice F is correct.* Tim's prediction is incorrect. The 10th flip of the coin is not related in any way to the first nine flips. The probability of the 10th flip landing on "tails" is the same as it was for every flip: 50%. The probability of the 10th flip landing on "heads" is also 50%. Because the two probabilities are equal, there is no greater chance of either happening.

MA.D.2.3.2

41 *A restaurant receives an order of 54 pounds of choice ribeye steaks priced at $8.75 per pound. The restaurant already owes $562.85 from a previous order. The restaurant manager agrees to pay for the current order as well as the previous balance owed. Which of the following shows the correct order of operations needed to determine the total bill?*

A. multiplication, addition
B. multiplication, division
C. addition, multiplication, addition
D. multiplication, subtraction, addition

Analysis: *Choice A is correct.* The manager should multiply $8.75 x 54 pounds, then add the amount owed from the previous order:

$$(\$8.75 \times 54) + \$562.85 = \$1,035.35$$

MA.E.1.3.3

42 *During the 56 years from 1950 through 2005, a total of 2,855 tornadoes have occurred in the State of Florida. The graph below shows the number of tornadoes that occurred each year.*

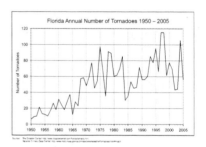

During which of the following years was the number of tornadoes closest to the mean number of tornadoes during this period?

F. 1960
G. 1970
H. 1980
I. 1990

Analysis: *Choice G is correct.* First, find the annual mean number of tornadoes for this 56-year period by dividing total number of tornadoes in the period by 56:

$$2,855 \div 56 \approx 51 \text{ tornadoes per year.}$$

Now, follow the vertical lines up for each of the four years until they cross the line graph, and then follow a horizontal path to the left to see how many tornadoes are associated with that year:

1960 – 30 or 31; 1970 – 48 or 49; 1980 – 60; 1990 – 56 or 57
The year 1970 is closest to the mean number of tornadoes for this 56-year period.

MA.E.1.3.3

43 *Which of the following statements cannot be supported with data from the graph?*

A. The total number of tornadoes in Florida varies greatly from year to year.

B. A particularly high number of tornadoes occurred in Florida in both 1997 and 1998.

C. Tornadoes in Florida are becoming increasingly more destructive.

D. The average number of annual tornadoes in Florida seems to be gradually increasing.

Analysis: *Choice C is correct.* There is no data in the graph about either the intensity of the storms or their destructive effects. It may very well be true that tornadoes in Florida are becoming increasingly more destructive, but there is no way to support this statement with this data.

MA.A.3.3.1

44 *Which of the following would need to be multiplied with*

$\dfrac{3}{4}$ *in order for the result to be 1?*

F. 0

G. 1

H. $\dfrac{3}{4}$

I. $\dfrac{4}{3}$

Analysis: *Choice I is correct.* Any fraction multiplied by its reciprocal equals one. For example,
$$3/4 \times 4/3 = 12/12 = 1.$$

MA.E.2.3.1

45 *Mark rolls a 6-sided die seven times. He has rolled a 6 twice, a 5 twice, and a 2 three times. What is the probability that he will roll another 2?*

A. 3:7

B. 1:6

C. 2:7

D. 1:3

Analysis: *Choice B is correct.* The probability of rolling any number on a die is 1 out of 6, because there are 6 different numbers that could appear. It doesn't matter how many times a number has been previously rolled.

MA.E.2.3.2

46 *Kara bought a bag of marbles. There are 6 red, 8 blue, and 14 green marbles in the bag. What are her odds for choosing a green marble at random from the bag?*

F. 1:2

G. 7:3

H. 1:1

I. 4:7

Analysis: *Choice H is correct.* Odds are not the same as probability. The odds for an event is defined as the ratio of the number of favorable outcomes to the number of unfavorable outcomes (in this case, the number of green marbles : the number of non-green marbles). There are 14 green marbles and 14 non-green marbles (6 red marbles + 8 blue marbles = 14 non-green marbles). Therefore, the odds for randomly choosing a green marble from the bag is 14:14 which reduces to 1:1.

MA.C.2.3.2

47 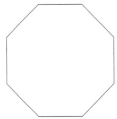 *The shape below will not tessellate by itself, but it will tessellate with one other regular polygon.*

How many sides must this other regular polygon have in order to tessellate with this octagon?

Analysis: *The correct answer is 4.* See the diagram below.

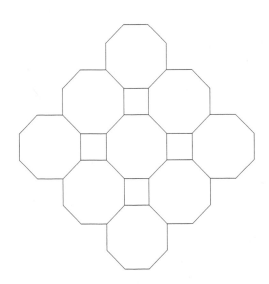

MA.A.1.3.2

48 *Which of the following choices puts the fractions in the correct order from the least to the greatest value?*

A. $\dfrac{1}{8}$, $\dfrac{3}{4}$, $\dfrac{5}{32}$, $\dfrac{3}{16}$

B. $\dfrac{3}{16}$, $\dfrac{1}{8}$, $\dfrac{3}{4}$, $\dfrac{5}{32}$

C. $\dfrac{1}{8}$, $\dfrac{5}{32}$, $\dfrac{3}{4}$, $\dfrac{3}{16}$

D. $\dfrac{1}{8}$, $\dfrac{5}{32}$, $\dfrac{3}{16}$, $\dfrac{3}{4}$

Analysis: *Choice D is correct.* First change all of the fractions to the same format, either decimals or fractions with the same denominator. For this analysis let's use fractions with the same denominator. The lowest common denominator of 4, 8, 16, and 32 is 32:

$$\frac{3}{4} = \frac{3}{4} \times \frac{8}{8} = \frac{24}{32}$$

$$\frac{1}{8} = \frac{1}{8} \times \frac{4}{4} = \frac{4}{32}$$

$$\frac{3}{16} = \frac{3}{16} \times \frac{2}{2} = \frac{6}{32}$$

and, of course, $\dfrac{5}{32}$ is already in the correct form.

Now it's easy to order the fractions from least to greatest just by looking at their numerators. The correct order is

$$\frac{4}{32} \quad \frac{5}{32} \quad \frac{6}{32} \quad \frac{24}{32} \quad \text{or} \quad \frac{1}{8} \quad \frac{5}{32} \quad \frac{3}{16} \quad \frac{3}{4} \text{, Choice D.}$$

MA.A.1.3.3

49 *What fractional part of $1,000.00 is $20.00?*

F. $\dfrac{1}{5}$

G. $\dfrac{1}{10}$

H. $\dfrac{1}{50}$

I. $\dfrac{1}{100}$

Analysis: *Choice H is correct.* First, set up the numbers as a fraction, but leave out the dollar signs and decimals in order to make things a bit easier, then reduce the fraction to lowest terms:

$$\frac{20}{1000} = \frac{1}{50} \text{, or Choice H.}$$

MA.B.2.3.2

50 *According to the Time Almanac 2005, the greatest rainfall ever recorded for a 12-month period occurred in Cherrapunji, India, from August 1860 to August 1861, when 1,042 inches of rain fell. Approximately, how many feet of rain is this to the nearest whole foot?*

Analysis: *The correct answer is approximately 87 feet.* Since there are 12 inches in a foot, divide 1,042 by 12:
$$1,042 \div 12 = 86.8333 \approx 87 \text{ feet.}$$

Mathematics Assessment Two—Correlation Chart

The Correlation Charts can be used by teachers to identify areas of improvement. When students miss a question, place an "X" in the corresponding box. A column with a large number of "Xs" shows more practice is needed with that particular standard.

Correlation	MA.A.1.3.1	MA.A.1.3.3	MA.A.4.3.1	MA.B.1.3.3	MA.B.1.3.1	MA.E.1.3.2	MA.D.1.3.2	MA.A.3.3.1	MA.A.3.3.2	MA.E.3.3.2	MA.E.1.3.1	MA.A.3.3.3	MA.D.1.3.1	MA.A.2.3.1	MA.A.1.3.4	MA.B.2.3.1	MA.B.2.3.2	MA.B.1.3.3	MA.E.3.3.1	MA.C.2.3.1
Answer	D	G	D	G	A	H	C	I	70.8%	C	H	4,800 years	D	H	C	H	D	I	D	I
Question	1	2	3	4	5	6	7	8	9	10	11	12	13	14	15	16	17	18	19	20

Student Names

Mathematics Assessment Two—Correlation Chart

Correlation		MA.B.1.3.2	MA.C.3.3.2	MA.C.1.3.1	MA.D.1.3.2	MA.D.2.3.1	MA.A.1.3.2	MA.A.5.3.1	MA.A.1.3.1	MA.B.1.3.4	MA.B.3.3.1	MA.B.1.3.1	MA.E.1.3.2	MA.A.5.3.1	MA.C.1.3.1	MA.A.3.3.3	MA.B.3.3.1	MA.C.2.3.1	MA.E.3.3.1	MA.D.1.3.1	MA.E.3.3.2
Answer	85°	A	I	C	G	B	29	G	D	H	C	G	D	G	D	H	C	H	B	F	
Question		21	22	23	24	25	26	27	28	29	30	31	32	33	34	35	36	37	38	39	40

Student Names

Mathematics Assessment Two—Correlation Chart

		MA.D.2.3.2	MA.E.1.3.3	MA.E.1.3.3	MA.A.3.3.1	MA.E.2.3.1	MA.E.2.3.2	MA.C.2.3.2	MA.A.1.3.2	MA.A.1.3.3	MA.B.2.3.2
Correlation											
Answer		A	G	C	I	B	H	4	D	H	87 feet
Question		41	42	43	44	45	46	47	48	49	50
Student Names											

Sunshine State Standards Checklist

The Sunshine State Standards Checklist can be used by teachers to easily identify what standard each question addresses in both Assessment One and Assessment Two.

	Question Numbers	
	Assessment One	Assessment Two
MA.A.1.3.1 (MC, GR)	15, 34	1, 28
MA.A.1.3.2 (MC)	23, 46	26, 48
MA.A.1.3.3 (MC, GR)	29	2, 49
MA.A.1.3.4 (MC, GR)	18	15
MA.A.2.3.1 (MC, GR)	31	14
MA.A.3.3.1 (MC)	28, 41	8, 44
MA.A.3.3.2 (MC, GR)	4	9
MA.A.3.3.3 (MC, GR)	2, 11, 40	12, 35
MA.A.4.3.1 (MC)	22, 43	3
MA.A.5.3.1 (MC, GR)	21	27, 33
MA.B.1.3.1 (MC, GR)	20, 47	5, 31
MA.B.1.3.2 (MC)	24, 35	21
MA.B.1.3.3 (MC, GR)	13	4, 18
MA.B.1.3.4 (MC, GR)	49	29
MA.B.2.3.1 (MC, GR)	42, 48	16
MA.B.2.3.2 (MC, GR)	1, 36	17, 50
MA.B.3.3.1 (MC)	25	30, 36
MA.C.1.3.1 (MC)	6, 17	23, 34
MA.C.2.3.1 (MC)	5	20, 37
MA.C.2.3.2 (MC)	33	47
MA.C.3.3.1 (MC)	50	
MA.C.3.3.2 (MC)	14	22
MA.D.1.3.1 (MC, GR)	19, 32	13, 39
MA.D.1.3.2 (MC, GR)	7	7, 24
MA.D.2.3.1 (MC)	30	25
MA.D.2.3.2 (MC, GR)	8, 37	41
MA.E.1.3.1 (MC, GR)	45	11
MA.E.1.3.2 (MC, GR)	27, 38	6, 32
MA.E.1.3.3 (MC, GR)	9	42, 43
MA.E.2.3.1 (MC)	12	45
MA.E.2.3.2 (MC)	10, 39	46
MA.E.3.3.1 (MC)	3, 16	19, 38
MA.E.3.3.2 (MC)	26, 44	10, 40